# Radiocarbon Dating New Zealand Prehistory Using Marine Shell

Matthew Schmidt

BAR International Series 842
2000

Published in 2019 by
BAR Publishing, Oxford

BAR International Series 842

*Radiocarbon Dating New Zealand Prehistory Using Marine Shell*



ISBN 9781841711300 paperback
ISBN 9781407351773 e-book

DOI https://doi.org/10.30861/9781841711300

A catalogue record for this book is available from the British Library

This book is available at www.barpublishing.com

BAR Publishing is the trading name of British Archaeological Reports (Oxford) Ltd. British Archaeological Reports was first incorporated in 1974 to publish the BAR Series, International and British. In 1992 Hadrian Books Ltd became part of the BAR group. This volume was originally published by John and Erica Hedges in conjunction with British Archaeological Reports (Oxford) Ltd / Hadrian Books Ltd, the Series principal publisher, in 2000. This present volume is published by BAR Publishing, 2019.

BAR
PUBLISHING

BAR titles are available from:

BAR Publishing
122 Banbury Rd, Oxford, OX2 7BP, UK
EMAIL info@barpublishing.com
PHONE +44 (0)1865 310431
FAX +44 (0)1865 316916
www.barpublishing.com

# CONTENTS

## ABSTRACT

Testing the reliability of marine shell for radiocarbon dating New Zealand prehistory has long been overdue. Though past studies have compared marine shell and charcoal radiocarbon dates from the same context to determine the reliability of marine shell, often the charcoal samples were unidentified. These studies were therefore invalid due to the risk of inbuilt age in the charcoal samples.

This dissertation investigates the reliability of marine shell for radiocarbon dating New Zealand prehistory. Paired charcoal and marine shell radiocarbon ages from archaeological sites excavated for this study and from already dated archaeological contexts, are compared. In this comparison, two issues are considered. First, factors which can affect the radiocarbon dating of marine shell and charcoal samples. Second, the effect of using either of two marine shell calibration methods in the comparison of marine shell and charcoal radiocarbon dates.

The upwelling and hardwater affects (causing shell ages to appear too old), and the incorporation of organic carbon (causing shell ages to appear too old or too young, depending on the source of the organics) are identified as important variables in dating shell. In the case of charcoal, it is found that although the identification of charcoal samples to twigs of short-lived species has been proposed as a solution to inbuilt age, this may still be a problem for charcoal samples. Incorrect data on the actual life span of a number of trees and shrubs identified as short-lived, and decay rates of various floral species after death may still influence charcoal dates.

A critique of the two methods available for calibrating marine shell radiocarbon ages is undertaken to examine the effect of each on age calibration, and in the comparison of marine shell and charcoal radiocarbon ages. The first method involves the application of a fixed reservoir correction (called *R*), to a marine shell radiocarbon age before calibration on an atmospheric calibration curve. The second method employs a modelled marine curve and a local reservoir value $\Delta R$, and is based on a box diffusion model which simulates fluctuations in R(t) through time. The two main differences identified between each calibration method are: a) a fixed *R* correction does not take into account changes in R(t) through time whereas the modelled marine curve does, and b) the modelled marine curve provides more defined calibrated age ranges compared to reservoir corrected calibrated marine shell dates due to the smoothness of this curve.

Paired charcoal and marine shell radiocarbon dates from 19 archaeological deposits excavated in this study are analysed. Of these 19 deposits, 14 possess charcoal and marine shell dates which are in chronological agreement after calibration. This is not dependent on the marine shell calibration method used. However, the modelled marine curve is found to produce more defined calibrated age ranges for the post-1500 AD period of New Zealand prehistory, compared to the expanded age ranges observed when using the reservoir correction calibration method. The modelled marine curve is argued to be both precise and accurate in calibrating shell radiocarbon dates for New Zealand prehistory based on four observations. First, the marine shell radiocarbon dates calibrated using the modelled marine curve are more often in agreement with paired calibrated charcoal dates, than marine shell ages calibrated using the reservoir correction calibration method. Second, the modelled marine curve successfully simulates changes in R(t) through time. Third, $\Delta R$ values from the archaeological sites analysed are in agreement with the current value for New Zealand. Fourth, shell dates calibrated using the modelled marine curve are in agreement with the archaeological evidence in dating the archaeological deposits.

Nine shellfish species commonly found in New Zealand archaeological sites are identified as reliable for radiocarbon dating. Though these species inhabited a variety of environments, this did not effect dating reproducibility between species. Even in locations where limestone strata is predominant, no hard water is detected. Charcoal ages are found to be less reproducible than marine shell for dating archaeological deposits. This is probably due to inbuilt age.

Five archaeological deposits where radiocarbon age differences are observed between charcoal and marine shell samples, are found to have been affected by post-depositional disturbance in prehistoric and modern times, and from the incorporation of sub-fossil marine shell.

An analysis of 40 previously dated archaeological sites with paired charcoal and marine shell radiocarbon dates is undertaken. Dubious radiocarbon ages are rejected using a discard protocol based on Anderson (1991), Spriggs and Anderson (1993) and Schmidt (1993, 1996a). Eight archaeological sites remaining after application of the discard protocol, show chronological agreement between the charcoal and marine shell dates, supporting the reliability of the shellfish species identified in this study, and use of the modelled marine curve.

In light of the results of the research, the time of colonisation of New Zealand is proposed based on archaeological sites with acceptable charcoal and marine shell radiocarbon chronologies.

**ACKNOWLEDGEMENTS**

I would like to thank the following people and organisations for their contributions and support in undertaking this research: Averille and Colin Schmidt, Rebecca Oskam, Maud Webber, Dr. Alan Hogg, Dr. Thomas Higham, Prof. Ken McKay, Prof. Emeritus Roger Green, Helen McKinnon, Dr. Fiona Petchey, Warren Gumley, Michael Taylor, Dr. Ian Barber, Dr. Rod Clough, Dr. Rod Wallace, Prof. Geoff Irwin, Dr. Thegn Ladefoged, Dr. Rod Wallace, Dr. Roger Sparks, Dr. Bruce McFadgen, Kevin Jones, Tony Walton, Dorothy Brown, Paul Speed, the Royal Society of New Zealand, the University of Waikato Radiocarbon Dating Laboratory, Institute of Nuclear Sciences, New Zealand Archaeological Consultants Ltd and all Iwi who provided permission to obtain samples from the sites for this study.

## LIST OF FIGURES

## LIST OF TABLES

# CHAPTER 1

# INTRODUCTION

*"The age of implements made by man has often to be determined, indeed in many cases can only be determined, by a consideration of the objects with which they have been discovered. Amongst these the most important are the teeth and bones of extinct mammals." Bertram C. A. Windle (1904 :45)*

## RELATIVE AND ABSOLUTE: DATING NEW ZEALAND PREHISTORY

The introduction of radiocarbon dating to archaeology by Libby and his colleagues in 1949 (Libby *et al.* 1949), was to affect the interpretation of prehistory worldwide. It meant that archaeologists would no longer have to rely upon relative dating techniques to order the past. In relative dating, the age of an archaeological site could only be derived through stratigraphy, and a comparison of artefact types. Where similar artefacts were found in sites great distances apart, it could be argued that the sites were occupied at a similar time in prehistory. These artefact types were then tied into historic records to develop calendar chronologies for sites and therefore regions.

Pottery was commonly used to construct relative prehistoric chronologies because distinctive patterns of manufacture were readily comparable between distant archaeological areas. This method, though, was confined to areas such as the Mediterranean, the Middle East, Europe, and the Americas, where pottery and ancient historic records could be used in parallel. In contexts where no written records existed, or where pottery was either non-existent or not commonly found in prehistoric sites, such as in Polynesia, the chronological placement of these sites in world prehistory depended on stratigraphy and artefacts, combined with local myth, oral tradition and genealogical chronologies (see Golson 1955:115,124). Archaeology was, therefore, restricted chronologically to material actually uncovered by the archaeologist or ancient historian which was distinctive of a certain calendar period.

In New Zealand, the earliest attempts to date archaeological sites were by Haast in the late 19th century. Haast believed that two different prehistoric groups had occupied the sites of Redcliffs and Shag River Mouth: the Moa-hunters, who were represented by cultural remains dominated by extinct avifauna, and the Shellfish eaters (or later Maori), who were represented by a predominance of marine fauna (see Haast 1875a, 1975b, 1875c). Haast suggested that the period of time separating these two cultural groups spanned many thousands of years based on the site's stratigraphy, artefacts, and the absence of moa in the upper layers. At the close of the 19th century, geological estimates of time were making way for ethnography to date archaeological finds, reducing the presence of people in New Zealand to within the last thousand years (see Anderson 1989:106-107). In the early 20th century, the incorporation of oral tradition to develop relative chronologies for archaeological finds was common, leading to the proposition of a number of colonisation and origins theories (see Barber 1995).

Possibly the most well known of these early theories was that proposed by Smith and Best (Smith 1921; Best 1914, 1916; Barber 1995:360). Smith and Best used oral tradition to argue that the first settlers in New Zealand were the Moriori (Maruiwi) from Melanesia. When New Zealand was colonised by Maori from Polynesia later in prehistory, the Moriori either fled to the Chatham Islands to avoid being subdued, or mixed with the 'superior' Maori people. An alternative to this theory was proposed by Skinner (1921, 1923), who emphasised the importance of the archaeological record. Skinner analysed cultural remains of Chatham Island Moriori and Southern New Zealand Moa-hunters. He showed that the material culture of Moriori and Maori were East Polynesian in origin, and there was no relationship between Moriori culture and Melanesia (Skinner 1921, 1923; Teviotdale 1932, 1938; Barber 1995:361-362).

When the first New Zealand archaeological material was radiocarbon dated in the 1950s, the principal questions concerned the date of first colonisation by Polynesians and the chronological placement of specific prehistoric events (Golson 1955:133-133). New Zealand was to the forefront of research on the new dating method and its use in archaeology (Golson 1955:130). In the mid 1950s, a radiocarbon dating committee was founded by the Nuclear Sciences Laboratory of the Department of Scientific and Industrial Research (DSIR) (now called the Institute of Geological and Nuclear Sciences (IGNS[1]) and New Zealand archaeologists (Leach pers. comm.[2]). The function of this committee was to select samples recovered by archaeologists for radiocarbon dating to ensure that the samples were associated with the archaeological event being dated. The committee also initiated a study on the reliability of various sample types for radiocarbon dating. One of these early studies by Lockerbie in 1957, examined variations in ages on moa bone, marine shell and charcoal from three occupation layers at the South Island moa-hunting site of Pounawea (Lockerbie 1959, 1980; see Chapter 3). Such studies, in collaboration with the screening committee, provided an initial base for the development of an absolute chronology for New Zealand prehistory.

### Radiocarbon Dating and the New Zealand Prehistoric Sequence

In 1959, Golson compared artefact types believed to be typical of the first colonisers of New Zealand from

[1] for this study, the abbreviation IGNS refers to the DSIR, INS, IGNS and Rafter laboratory.

[2] Dr. B.F. Leach, Museum of New Zealand, Wellington.

radiocarbon dated archaeological sites (such as Wairau Bar Moa-hunting site in the South Island), with artefacts from sites thought to be typical of later Maori settlement and whose forms were still being used by Maori when Cook arrived (such as found at Oruarangi in the North Island) (Golson 1959:29-74). His interpretation represented a turning point in New Zealand archaeology. Golson proposed the use of the terms Archaic and Classic to describe the prehistoric sequence of New Zealand, with the radiocarbon dates available providing a *terminus ante quem* for the sequence (which previously could only be obtained through oral tradition) (Golson 1959:36, 47). The artefacts now became markers in relative chronologies for prehistoric New Zealand supported by the radiocarbon dating technique. This meant regional chronologies for New Zealand prehistory could now be constructed independent of genealogy or oral tradition, and that early sites were not simply those which contained moa remains (Golson 1959:36).

Golson (1959), therefore, had not only produced a workable cultural sequence for New Zealand prehistory (which he recognised as being flexible as more data was made available), but he emphasised the importance of the radiocarbon chronology in the development of his Archaic and Classic Phases, alongside the artifactual evidence (Golson 1986; Groube 1993:6-17). However, as Golson and other New Zealand archaeologists soon found, the radiocarbon dating technique was as new to science as it was to archaeology.

From 1959 to the early 1970s, archaeologists were struggling to place archaeological sites into either the Archaic or Classic Phases of New Zealand prehistory, when the radiocarbon chronologies from their sites were considered alongside artefact types thought to be typical of certain phases (see Chapter 3). This quandary brought about conflict between Golson's cultural sequence based on artefact typologies and radiocarbon data, and relative ages for sites based on artefact assemblages and oral and traditional accounts (see Duff 1962:205-209). One archaeological site which conflicted with Golson's cultural sequence was the Takehe Valley site, which was omitted by Golson (1959) in the development of his cultural chronology.

In 1949, while undertaking a field survey in Takahe Valley in the South Island, K. Meirs and R.A. Falla found feathers of a variety of native birds protruding from the dust floor of a limestone cave (later to be called Shelter A). Investigation uncovered scorched vertebrae of the small bush moa *Megalapteryx* as well as a section of pelvis with evidence of cut marks (suggested on examination by Skinner as being caused by a metal knife) (Duff 1956:76-79). Found alongside the *Megalapteryx* bones and other avian fauna remains on a bedding of tussock, was a sandal made from plaited flax, flax nooses, a bird lure made from weka feathers, and a pointed fire-stick (Duff 1956:76). Duff later excavated Shelter A in 1950 and 1951 to confirm the stratigraphy of the site, and found flakes from an adze and a large flake tool. He saw that the deposit was only 12.7cm in depth, confined to a square metre, and therefore probably represented either a single visit of a large hunting party, or brief visits by a smaller group over a few years (Duff 1956:77-79).

In interpreting the chronology of the site, Duff doubted that the cut marks on the moa pelvis were from a metal knife, but believed that the site was post-fleet in origin and was possibly occupied as late as the early 1800s (Duff 1956:80). When considering the radiocarbon evidence, Duff concluded:

*"The limitations of the radiocarbon method for the immediate past are seen in Rafter's attribution of a sample of tussock bedding from Shelter A, Takahe Valley, to 110 ± 60 years from 1955 (from A.D. 1785 to 1905). Assuming 1905 to be impossible, we find that the earlier date does support the hypothesis of an eighteenth-century migration advance on other grounds. Indeed, it strengthens Skinner's proposition that this occupation was indeed post-European. The anomaly so frequently encountered in radiocarbon analysis is illustrated for this site by Dr. Rafter's determination of strips of totara bark, obviously from the same cultural level as the tussock, as 700 ± 60 years from 1955 (A.D. 1195 to 1315). It is a fair assumption here, however, that the bark was stripped from already aged or dead trees encountered on the shores of Lake Te Anau." (Duff 1956: XII).*

Takehe Valley illustrated the problems associated with the use of radiocarbon during this period. These problems came from two areas: 1) provenance of the samples obtained by archaeologists for dating sites, and 2) the reliability of charcoal, marine shell and bone for radiocarbon dating archaeological contexts (see Chapter 3).

At Takehe Valley, for example, the exact provenance of the samples was dubious. Also, the site was openly exposed to the environment (the Shelter A description implying a cave situation which the site was not) and no post-depositional disturbance of the site had been taken into account by Duff. Because the radiocarbon dates from the site were in conflict, Duff had selected the radiocarbon age which was important to his own idea of the New Zealand prehistoric sequence, and supported this using post-Fleet oral tradition[3].

The radiocarbon chronology of Ngaroto swamp *pa* was also hindered by unreliable provenancing of dated material (Schmidt 1993). Excavated during the late 1960s, Ngaroto was the deepest stratified archaeological site in New Zealand at the time, with cultural material spanning a depth of 4 meters (Shawcross 1968). It was also the *pa* site where the famous Maori carving Uenuku was found in 1906 (Best and Anderson 1926:70). Because of poor provenancing of the dated charcoal and marine shell samples, it was unclear whether occupation of the *pa* began either 800 or 250 years ago, as it could not be determined which sample came from the basal layer, and which came from one of the upper layers (Schmidt 1993). Otakanini *pa* was another example in which unreliable provenancing of the material dated implied a prehistoric site was constructed in post-contact times. The excavators of the *pa* wished to date the various phases of fortification, and so took samples from preserved palisade posts found during excavation. But wood thought to be a Period III palisade post was found after radiocarbon dating to have derived from a modern fence post, thus not only throwing into doubt this particular radiocarbon sample, but the remainder of the samples selected as well (Bellwood 1973:173).

[3] Later analysis of the Takahe Valley site and further dating by O'Regan (1992:172-176), derived radiocarbon ages of (NZA-2228) 611 ± 51 years BP on feathers, tussock and beech leaves combined, and (NZA-2227) 628 ± 39 years BP on a small piece of bone from the butchered *Megalapteryx* ischial process.

The anomalous radiocarbon chronologies from the Takehe Valley moa-hunting site and Ngaroto and Otakanini *pa*, were, therefore, due to dubious sample provenancing, and the results were not the fault of the radiocarbon method itself. Though scientific excavation techniques were introduced to New Zealand during the mid 1950s, some archaeologists were not considering the implications of unreliable provenancing of samples taken for radiocarbon dating (Golson 1955:349; Golson 1986; Groube 1993:6-17).

In many cases, despite radiocarbon ages having accurate provenance, dates were found to be too old because of the material being dated (see Chapter 3 and 4). This situation occurred at Tiromoana *pa* excavated by Fox from 1974 to 1975 (Fox 1978). Here, the radiocarbon ages on charcoal and wood from pit structures stretched back to 3530 ± 90 years BP (NZ-1913), with one radiocarbon age of 1050 ± 90 years BP (NZ-1914) from the post of a pit accepted as the earliest evidence of agriculture in the North Island (Fox 1978:1,28,50). The posts did come from the provenance described, but the timbers produced greatly varying ages because the palisades used in prehistory came from ancient totara logs. The 1050 ± 90 years BP (NZ-1914) age was accepted because at the time New Zealand was believed to have been settled 1000 years ago (see Davidson 1984:123-125).

The problem of sample provenance could only be solved through good excavation technique, and so was dependent on the excavation skill of the archaeologist. The reliability of various sample types for dating, however, was problematic. As archaeologists were advised by the New Zealand Geological Survey that charcoal and wood were the most preferable material for dating (Buist 1962:249), the majority of radiocarbon dates from archaeological sites prior to 1976 (see below) comprised these sample types (see Anderson 1989; Schmidt 1993, 1996a). Hence, many of the problems with radiocarbon chronologies from sites were attributed to these materials. Although dates on charcoal or wood which were significantly older than expected were proposed as being due to relic wood, such as at Tiromoana, archaeologists were finding it difficult to ascertain which ages from some sites were too old, and which ages reflected the first arrival of people in New Zealand. For example, one archaeological site of vital importance to New Zealand prehistory is Tairua on the Coromandel Peninsula, excavated by Green and Smart in the early 1960s. This site contained a pearl shell lure shank of East Polynesian origin in Bed 2 (Green 1967). Though moa bone in the deposit indicated that the occupation was early, archaeologists at the time were unable to ascertain whether it represented an early or late colonisation episode. This was because the two charcoal radiocarbon ages from the deposit conflicted, deriving dates of 878 ± 49 (NZ-594) and 443 ± 40 (NZ-595) years BP (Rowland 1976:6)[4].

The combination of old and young charcoal radiocarbon ages was thus disrupting the division of the Archaic and Classic Phases based on archaeological artefacts and features, pushing back the appearance of some sites, such as *pa*, to the Archaic, and the existence of Archaic fauna such as moa, into the Classic. The relic wood problem was also causing sample type studies, such as those undertaken by Lockerbie (1959, 1980), to become problematic (see Chapter 3). Questions about the reliability of sample types being dated could not be answered with confidence. Charcoal, wood, marine shell and bone samples which were radiocarbon dated from the same provenance sometimes showed varying ages, and though the bone or marine shell dates, for example, may have been consistently correct, this conclusion could not be justified due to the varying results of charcoal ages from same context. These issues were additionally confounded by the short prehistoric time span for New Zealand which required a reliable sample type for radiocarbon dating prehistoric events within the last 1000 years.

Cultural sequences for archaeological sites were therefore dependant on whether the radiocarbon ages were being accepted or rejected by archaeologists. Often observations were made that charcoal dates were too old because of relic wood, but archaeologists continued to select charcoal and wood samples without consideration of inbuilt age[5] until after dating had been undertaken (see Trotter 1968; Law 1971; Rafter *et al.* 1972; see Chapter 3).

Other dating techniques were often used to aid in interpreting radiocarbon chronologies. Some archaeologists incorporated oral tradition or historic evidence as a means of either justifying or omitting radiocarbon evidence (Duff 1956, 1961; Trotter 1967a:141, 1967b; Fox 1978). Artefact types again became of central importance in providing relative ages for sites, such as Duff's (1959) adze types, or Hjarno's (1967) fish hook chronology, as they had before radiocarbon dating (see Green and Shawcross 1961:210-213; Green 1962:8-16). The possible age of *pa* were relatively dated by a belief that more complex forms were later developments in New Zealand prehistory (Groube 1970; Schmidt 1993). Identification of the Kaharoa Ash, then dated to *ca.* 900 years BP (Pullar 1959, 1962), the Rangitoto Ash fall of *ca.* 600 years BP (Law 1975), or the presence of Loisel's pumice (Law 1975:49) were other chronological markers also employed. Obsidian hydration was investigated as a technique for dating sites, but was not successful due to issues such as the effect of temperature and obsidian composition on hydration rate (Ambrose and Green 1962:247-248; Green 1962). Fluorine analysis of bone was attempted but was found not to be accurate enough for New Zealand prehistory (Trotter and Malthus 1967:151-157).

New Zealand archaeologists were therefore left with few methods for accurately dating prehistoric deposits outside of radiocarbon dating because of the unavailability of other absolute dating methods. The few solutions suggested by various researchers to counter problems associated with the dating of charcoal, marine shell or bone were of little help (see Chapter 3). One method occasionally implemented was to obtain wood and charcoal samples which had tree rings still present. It was believed that the number of rings found in the charcoal branch, for example, could provide an

[4] Schmidt and Higham (1998) have now redated the Tairua site using reliable shellfish species identified in this study. The redating suggests the pearl shell lure arrived in New Zealand by the late 13th or early 14th century AD.

[5] Inbuilt age may be defined as "the difference in age between the death of the sample and the archaeological event dated. For wood, it is the combination of growth age (the age of old wood in a tree) and storage age (the time the tree was lying around before it was used)." (McFadgen *et al.* 1994:223).

estimate of how many years of tree growth had gone by before the wood was used in prehistory. An estimate was also made of how many of the outer rings may have decayed. Using this method, wood or charcoal from the outermost tree rings were dated and an error attributed to the radiocarbon age for the missing rings. An example of this practice was the dating of Warehou *pa* by Brodie (1962:157-159). Wood from a palisade was identified as totara, the rings counted (90 in total), and a date taken from the 60th ring. The radiocarbon age was found to be 310 ± 60 years BP (NZ-?), Brodie (1962:157) noting that growth of the timber after the 60th ring could increase the age of the site between 60 and 100 years (this method was also used at Tiromoana *pa* by Fox (1978)).

Only a handful of researchers suggested the selection of twigs when choosing charcoal for dating, but the importance of species selection was not discussed (Law 1971; Grant-Taylor 1974; see Chapter 3). Archaeologists were advised when dating marine shell to avoid dating mudsnail (*Amphibola crenata*), but experiments on reliability largely focused on dating the carbonate or keratin fractions of shell, rather than rigorous investigation of reliable shellfish species (Trotter 1967a:138).

However, in 1968, the archaeologist B.F. Leach suggested to the New Zealand Archaeological Association that wider screening of samples submitted for dating archaeological sites was required so reliable materials could be isolated (Leach pers. comm.). Then in 1976, a solution to the problems of dating charcoal samples was formulated by Leach and a botanist, B. Molloy (Leach pers. comm.; see Chapter 3; Anderson and McGovern-Wilson 1990; Anderson 1991). The solution was to inform archaeologists that all charcoal and wood samples submitted for radiocarbon dating should be identified to twigs of short-lived species, so as to reduce the risk of inbuilt age. Since this time, archaeologists have sent charcoal and wood samples to be identified prior to dating. This was initially undertaken by B. Molloy of the University of Canterbury, and now by R. Wallace of the University of Auckland Department of Anthropology. This solution has significantly improved the accuracy of charcoal ages (see Higham 1993), and charcoal identified using this initial screening process is believed by many archaeologists today as providing samples for dating which should yield accurate ages (though debate still continues over charcoal's reliability) (see Anderson 1991; Chapter 3).

But what about dating archaeological sites where charcoal and wood samples are absent, or where it is difficult to obtain datable charcoal? In this regard, archaeologists since 1976 have either considered alternative carbon-containing materials to radiocarbon date, or sought other absolute chronological methods to determine the age of a site. The use of obsidian hydration has again been addressed for dating, but development of this technique is at an early stage in New Zealand (see Sutton and Shepard 1994; Shepard *et al.* 1996; Stevenson *et al.* 1996), and radiocarbon ages are still required as a check on the obsidian hydration chronology. The Kaharoa Ash has now been firmly dated to 665 ± 20 years BP (Higham pers. comm.[6]) and can be used effectively as a chronological marker for New Zealand prehistory. However, its deposition is location specific, and it provides only a relative age for sites found above, or below it.

The only alternatives remaining for a prehistorian wishing to accurately date an archaeological deposit when charcoal for radiocarbon dating is absent, is to radiocarbon date marine shell or bone. The difficulty is, however, that the reliability of such sample types for dating New Zealand prehistory has never been systematically and rigorously investigated. Certainly, marine shell has been generally accepted by archaeologists and radiocarbon laboratories as reliable, but debate has continued over its accuracy in some contexts (see Law 1971; Rafter *et al.* 1972; Grant-Taylor 1974; Law 1981; Higham 1993; Chapter 3). The reliability of New Zealand archaeological bone for dating is at present poorly understood, though research in the dating of fish bone and rat bone has recently been undertaken (Redvers-Newton and Coote 1994; Redvers-Newton 1995; Petchey 1998). Petchey's (1998) research has now identified the fish species *Pagrus auratus* and *Thyrsites atun* as reliable for radiocarbon dating.

The reliability of marine shell is probably of more concern to archaeologists than bone in view of its frequent use for radiocarbon dating New Zealand prehistory since the mid-1950s. This is especially true in the dating of particular archaeological sites such as *pa* which often lack datable charcoal. For *pa*, 55% of radiocarbon dates are from marine shell, with 90% of *pa* dated since 1976 using this sample type (Schmidt 1993, 1996a). In addition, the most common material from New Zealand archaeological sites dated by the University of Waikato Radiocarbon Dating Laboratory has been marine shell, making up 79% of prehistoric dates (Higham pers. comm.).

## AIMS OF THIS STUDY

The aim of this study is to investigate the use of marine shell for radiocarbon dating New Zealand prehistory. This study will therefore evaluate:

1) the reproducibility of radiocarbon dates between shellfish species commonly found in New Zealand archaeological sites. It is possible that some marine shell species will not produce consistently reliable radiocarbon ages. By identifying reliable shellfish species to date, New Zealand archaeologists will be able to confidently select particular shellfish species for dating purposes;
2) the chronological agreement between calibrated marine shell and charcoal radiocarbon dates from contemporary archaeological deposits. Both marine shell and charcoal radiocarbon ages from a contemporary archaeological deposit should agree in dating a prehistoric event. However, this may be dependant on suitable shellfish species dated, inbuilt age of charcoal, or the marine shell calibration method employed (see Chapters 4 and 5);
3) variability in charcoal ages. Charcoal radiocarbon ages may be less reproducible than marine shell dates (due to factors such as inbuilt age), and so less desirable to chose as a dating medium for the archaeologist;
4) the effect of hard water on the dating of New Zealand archaeological marine shells. Strata containing ancient carbon are extensive in New Zealand, therefore the

[6] Dr. T.F.G. Higham, University of Waikato Radiocarbon Dating Laboratory.

hardwater effect is a concern to prehistorians when dating archaeological sites in such areas;

5) the applicability of the Stuiver and Becker (1993) modelled marine calibration curve to New Zealand prehistory. There are two methods by which marine shell radiocarbon dates may be calibrated (see Chapter 5). An investigation of the applicability of the Stuiver and Becker (1993) modelled marine curve is important for prehistorians, as choice of this calibration method may effect the accuracy of dating an archaeological deposit;

6) the current ΔR value for New Zealand (Higham and Hogg 1995). The modelled marine calibration curve depends upon the local reservoir correction (ΔR) being non-time dependant (see Chapter 5). If it is shown from this research that ΔR may vary through time, then use of the current value for calibrating all New Zealand prehistoric marine shell ages may be invalid.

The investigation of these variables will have implications for certain critical issues in New Zealand archaeology, such as the date of colonisation based on archaeological sites with both charcoal and marine shell radiocarbon chronologies.

Understanding the principles behind radiocarbon dating is important in undertaking this investigation. This is the subject of Chapter 2 of this study. This chapter also explains the radiocarbon dating technique used at the University of Waikato laboratory, as well as statistical and calibration methods employed in the analysis of archaeological radiocarbon ages.

Chapter 3 comprises a historical review of the opinions of archaeologists and radiocarbon dating researchers on the choice of materials for radiocarbon dating New Zealand prehistory. Many views on this issue are presented, as well as studies undertaken over the years by archaeologists and radiocarbon dating laboratories comparing ages between differing sample types. A solution to this sample type issue is proposed.

Chapter 4 identifies problems associated with the radiocarbon dating of charcoal and marine shell which may cause discrepancies between radiocarbon ages on these two sample types. In Chapter 5, both methods by which marine shell radiocarbon ages can be calibrated to years AD are explained and scrutinised. This is undertaken to ascertain whether either marine shell calibration method affects the placement of an archaeological event in calendar time.

Chapter 6 describes the archaeological sites excavated for this dissertation to obtain charcoal/marine shell pairs. The sites were chosen to provide a cross-section of the marine and terrestrial environments associated with archaeological sites around New Zealand. They were also selected according to the large variety of shellfish species present, to enable variations in radiocarbon dates on particular shellfish species to be investigated. Data on the artifactual and faunal material recovered, stratigraphy, site history, surrounding geology, and the environment associated with the sites during their formation is presented. This information is vital as the effect of environment and habitats of New Zealand shellfish species on the dating of marine shell have seldom been rigorously investigated. Reconstruction of the vegetation type surrounding the archaeological sites during occupation is also examined to provide information regarding the fuel sources which were available in prehistory. This information may explain variations in charcoal radiocarbon ages due to factors such as inbuilt age.

In Chapter 7, an analysis of the paired radiocarbon ages from the sites excavated for this dissertation is undertaken. Archaeological sites where calibrated marine shell and charcoal radiocarbon dates are in agreement are discussed, and reasons for discrepancies between marine shell and radiocarbon ages from other sites are proposed. In Chapter 8, the overall results from the study are evaluated, drawing conclusions on the use of marine shell for radiocarbon dating New Zealand prehistory, the advantages and disadvantages of using shell compared to charcoal, and the effect of using either of the currently employed marine calibration curves.

# CHAPTER 2

# THE RADIOCARBON METHOD

*"Potentially more sensitive is a new method of dating organic material by means of its radiocarbon content. This method, a by-product of atomic research,...may eventually enable specimens up to 40,000 years old to be dated objectively." Sir Mortimer Wheeler (1954:50)*

## INTRODUCTION

This chapter explains the basis of the radiocarbon dating method by which absolute dates have been obtained from archaeological sites throughout the world. The radiocarbon dating technique employed at the University of Waikato Radiocarbon Dating Laboratory to date New Zealand prehistory is also described. Additionally, the statistical methodology and calibration procedures used in this study are presented.

## PRINCIPLES AND ASSUMPTIONS

Carbon occurs in nature as three isotopes: $^{12}C$, $^{13}C$ and $^{14}C$. Proportionally, $^{12}C$ makes up *ca.* 99% of carbon, $^{13}C$ *ca.* 1% and $^{14}C$ *ca.* 1 x $10^{-10}$%. $^{14}C$ is formed continually in the lower stratosphere and upper troposphere by the interaction of neutrons produced by cosmic rays with nitrogen atoms (Bowman 1990):

$$^{14}N + {}^{1}n \rightarrow {}^{14}C + {}^{1}H$$

The $^{14}C$ produced is rapidly oxidised to form carbon dioxide ($^{14}CO_2$). The $^{14}CO_2$ is then taken in by plants through photosynthesis where it enters the terrestrial food chain, and by direct exchange with the ocean where it is incorporated into oceanic organisms (Figure 2.1). After production, $^{14}C$ begins to decay according to the process:

$$^{14}C \rightarrow {}^{14}N + \beta^-$$

where β is a beta particle emitted as a product of radioactive decay. When the production and decay of $^{14}C$ is in equilibrium, there is a constant concentration of $^{14}C$ in each carbon reservoir and thus in the associated living organisms (Bowman 1990:10; Aitken 1990:57-58). When an organism dies, it no longer exchanges carbon with its particular reservoir and therefore the remaining $^{14}C$ in the plant or animal decays without replenishment according to a known rate. The rate of radioactive decay is called the half-life. For $^{14}C$, this means that after 5568 years, half of the original amount of $^{14}C$ would remain in a dead organism (Figure 2.2).

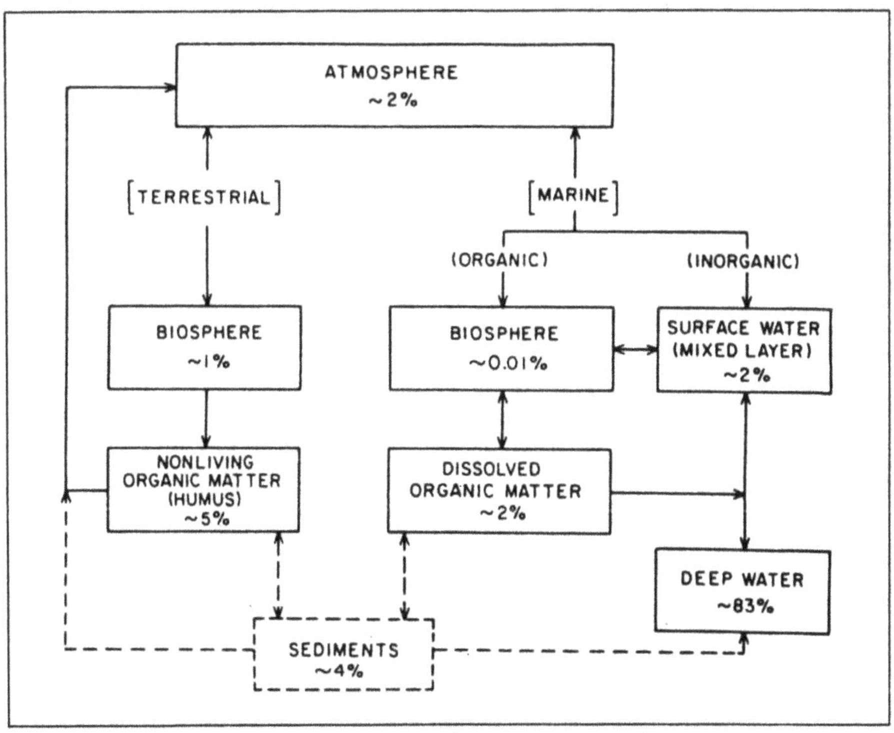


Figure 2.1. Carbon reservoirs in nature (%=%Carbon) (from Taylor 1987:8).

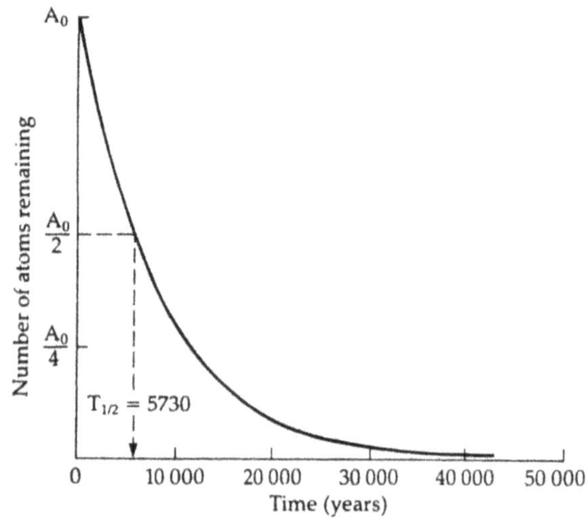


1 The decay of a radiocative element follows the exponential decay law. The primary characteristic of exponential decay is that the percentage decrease in number of atoms per unit time is constant; hence after each half-life the number of atoms remaining is halved: if there are $A_0$ atoms to begin with, then after one half-life there will be $A_0/2$ atoms remaining; after two half-lives, $A_0/4$ remain; after three, $A_0/8$ and so on.

Figure 2.2. The half-life of $^{14}C$ (from Bowman 1990:11).

The half-life of $^{14}C$ was originally calculated by Libby *et al.* (1949) as 5568 years (known as the "Libby" half-life).

Later measurements increased this to 5730 years (called the "Cambridge" or New Half-Life) (Taylor 1987:9). After 50 - 60K years, very little $^{14}C$ remains, and therefore this becomes a maximum age to which this chronological technique extends. (In practice, however, 40K years BP is seldom exceeded in archaeology (see Allen and Holdaway 1995:101-112)). It is from this process of decay and knowledge of the half-life of $^{14}C$, that we may find the age of an organism.

Using $^{14}C$ as a dating technique depends on a number of assumptions. These are (Taylor 1987:3):

"1. The concentration of $^{14}C$ in each carbon reservoir has remained essentially constant over the $^{14}C$ time scale;

2. there has been complete and rapid mixing of $^{14}C$ throughout the various carbon reservoirs on a worldwide basis;

3. carbon isotope ratios in samples have not been altered except by $^{14}C$ decay since these sample materials ceased to be an active part of one of the carbon reservoirs (as at the death of an organism);

4. the half-life of $^{14}C$ is accurately known; and

5. natural levels of $^{14}C$ can be measured to appropriate levels of accuracy and precision."

The first test of the radiocarbon method was carried out by Arnold and Libby in 1949 (Aitken 1990:58). Here radiocarbon ages of wood taken from Egyptian tombs of known age were compared. Many of the radiocarbon and historic ages were in close agreement illustrating that the radiocarbon dating method was a very accurate and precise tool for archaeological research. However, it was observed early on in radiocarbon dating that the production and movement of $^{14}C$ in the atmosphere have not been constant through time (de Vires 1958; Willis *et al.* 1960). Because of this, radiocarbon dates could not always be directly converted to calendar ages.

Calibration curves were therefore developed in the early 1970s to derive calendar ages from radiocarbon dates (see Suess 1967, 1970; Damon *et al.* 1972; Aitken 1990:3). Because tree rings are representative of events of known historic age, radiocarbon measurements of successive rings allowed for correction of fluctuating $^{14}C$ levels through time. This process thus produced a calibration curve to convert a radiocarbon age directly to a calendar date (Figure 2.3). These calibration curves were applicable for radiocarbon dates derived from charcoal and wood samples, but not for samples formed in the ocean such as marine shell. It was found that marine shell radiocarbon ages were *ca.* 400 years older than charcoal or wood radiocarbon dates of the same calendar age (see Chapter 5). This was because the ocean reservoir experiences slow mixing and absorption of atmospheric $^{14}C$. Therefore, before radiocarbon dates on marine shell could be applied to a calibration curve to derive a calendar age, *ca.* 400 years first had to be subtracted. This age difference was termed the Reservoir Correction or *R* (see Chapter 5).

## RADIOCARBON DATING AT THE UNIVERSITY OF WAIKATO LABORATORY

### Sample Pretreatment

Higham (1993:98-119) undertook extensive studies on the pretreatment of New Zealand archaeological samples prior to radiocarbon dating. He concluded that, "At the level of precision used in the examples given above, contaminants are either negligible in their impact, or their side effects are being masked by higher standard errors. Given the situation encountered at Pleasant River it is wise to err on the side of caution and pretreat all archaeological samples with simple routine procedures." (Higham 1993:119). For this study, pretreatment procedures were developed for charcoal and marine shell following those of Higham (1993:98-119; pers. comm.), including the removal of physical and chemical contaminants and, for marine shell, the use of X-ray powder diffraction (XRD) analysis to test for recrystallisation.

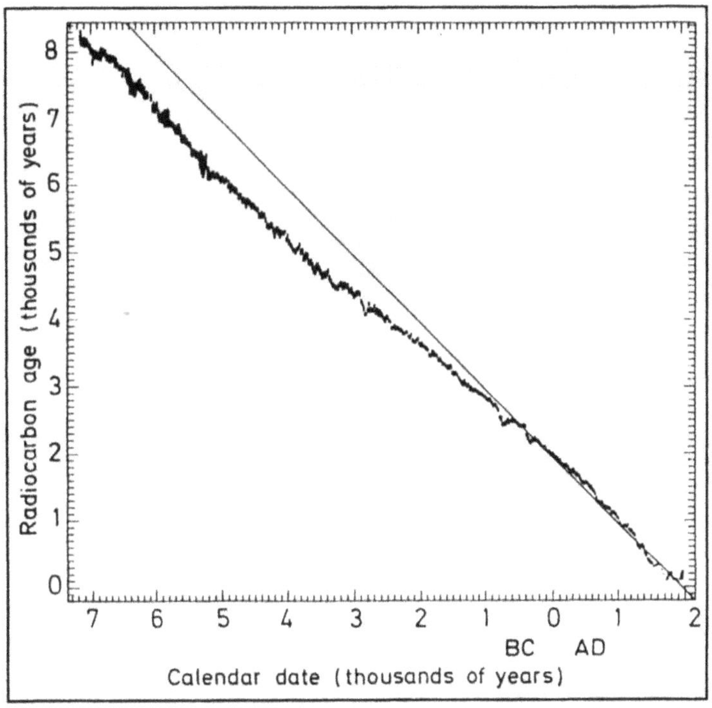


Figure 2.3. Relationship between radiocarbon ages and calendar dates derived through the dating of tree rings. Curves such as this are used in the conversion of a radiocarbon age to a calendar date (from Aitken 1990:99).

Contaminants of concern for charcoal samples may be derived from pre- and post-sampling sources. Contamination after deposition includes root infiltration from plants growing above the site, humic acid absorption from upper level decomposed plant remains intruding down onto the sample, and adsorption of carbonates from ground water flowing through the site (Hogg 1982:18; Higham 1993:103). The first two contaminants will make the charcoal appear too young when dated, whilst carbonate contamination can cause charcoal ages to appear too old. Sources of contaminants after sample recovery can be from packaging of the sample using paper products, from hair, cigarette ash (making the dates more modern), and oil or grease (making the dates too old) (Hogg 1982:21).

As mentioned above, recrystallisation of shell after deposition in an archaeological context must be tested for, even though with the short length of New Zealand prehistory this situation seems unlikely. Higham has noted that, "No New Zealand archaeological shells have ever been identified as contaminated by recrystallisation, so XRD [X-ray diffraction analysis] is not routinely implemented"[7] (Higham 1993:99). Recrystallisation of marine shell involves the dissolution of the shell and recrystallisation of calcium carbonate in the form of calcite. This process can only be detected in aragonite-secreting organisms. It causes a radiocarbon age to be too young, as the recrystallisation process incorporates modern carbon into the shell structure[8] (Hogg 1982:26; Gupta and Polach 1985:11, 128). Aragonite secreting shellfish species typically used in radiocarbon dating New Zealand prehistory include *Paphies australis* (pipi), *Paphies subtriangulatum* (tuatua), *Austrovenus stutchburyi* (cockle), *Cominella* sp. (whelks), *Venerupis largillerti* (oblong cockle), *Mytilus edulis aoteanus* (blue mussel), *Perna canalicula* (green mussel) and *Lunella smaragda* (catseye). An XRD analysis of these species will typically show the presence of aragonite only, but if calcite is detected, recrystallisation has occurred and the marine shell is rejected for dating. Calcite secreting organisms such as *Crassostrea glomerata* (rock oyster) are not X-rayed as recrystallisation is undetectable.

Contamination by older carbon is more serious than contamination by modern carbon for New Zealand archaeological materials because of their young age (Table 2.1). Caughley (1988:246) notes that for a sample 900 years old, contamination by 1% modern carbon will reduce the true age of the sample by approximately the same percentage. However, for every 1% of old carbon contamination, there is a *ca.* 9% error (*ca.* 80 years). Therefore for New Zealand prehistory, we should be more concerned about contamination from old carbon sources than from modern.

## Charcoal Pretreatment

Though Higham (1993:119) found that treated and untreated charcoal samples from his sites showed no significant difference in their true radiocarbon age, a routine pretreatment procedure was undertaken for this dissertation

[7] Anderson *et al.* (1996:66) have noted that a sample of paua (*Haliotis iris*) from the Shag Mouth moa-hunting site was possibly affected by recrystallisation.

[8] Gupta and Polach (1985:128) note that under particular environmental conditions, recrystallisation may sometimes cause shell ages to appear too old.

| | % contamination with MODERN carbon | | | | | |
|---|---|---|---|---|---|---|
| | 0% | 1% | 5% | 10% | 25% | 50% |
| **Est Age BP.** | 900 | 890 | 850 | 810 | 670 | 440 |
| **% error** | 0 | 1 | 5 | 11 | 26 | 51 |
| | | | | | | |
| | **% contamination with OLD carbon** | | | | | |
| | 0% | 1% | 5% | 10% | 25% | 50% |
| **Est Age BP.** | 900 | 980 | 1320 | 1770 | 3280 | 6630 |
| **% error** | 0 | 9 | 47 | 97 | 264 | 637 |

Table 2.1. The effect on a 900 years old sample when contaminated by differing amounts of modern and old carbon (from Caughley 1988:247).

to remove obvious contaminants such as root hairs and carbonates from charcoal samples.

The charcoal sample was first screened for physical contaminants such as root hairs, sand and pieces of shell. The sample was then carefully crushed using a mortar and pestle to reveal any additional contaminants missed during initial screening, and to expose as much of the surface area as possible to chemical pretreatment. The charcoal was then washed with hot 10% HCl for 24 hours to remove carbonates, then rinsed with deionised water to achieve a neutral pH level, and again screened for any physical contaminants. The charcoal was then dried in an oven at 80°C.

## Marine Shell Pretreatment

Before a marine shell sample could be dated, I tested each shell for recrystallisation by XRD using a Philips PW 1729 X-ray generator. The shell sample was first cleaned in an ultrasonic bath for 15 minutes to remove any adhered debris, and then left in an oven overnight to dry. A small portion of shell was then removed from each shell in the sample, the pieces combined, and then ground using a mortar and pestle. The ground shell was then X-rayed. No aragonitic marine shells radiocarbon dated in this study showed evidence of recrystallisation.

## Radiocarbon Dating by Liquid Scintillation Counting of Benzene

There have been four techniques by which the radioactive decay of $^{14}C$ has been measured to determine the age of archaeological and geological samples. The first was the use of a screen walled Solid Carbon Counter by Libby, Arnold and Anderson at the close of the 1940s (Taylor 1987:76, 82). Here elementary carbon was coated onto the inside of a metal container and a Geiger counter placed inside. The Geiger counter measured the number of beta particles radiating from the sample over a given period of time.

During the early 1950s, two further techniques were developed which provided increased precision in $^{14}C$ decay measurement. Gas Counting involved the conversion of the sample into $CO_2$ for the measuring of $^{14}C$ activity. In Liquid Scintillation (LS) Counting, a sample was converted into benzene ($C_6H_6$), and then a scintillator added. A scintillator is a compound which reacts fluorescently to the emission of radioactive particles by producing short bursts of photons (Taylor 1988:86-88; Polach 1992:199). Because of this ability, the emission of β particles through the radioactive decay of $^{14}C$ can be detected and therefore counted to determine $^{14}C$ activity. Benzene was ideal for LS counting because of its high fraction of carbon and excellent light transmission qualities. LS Counting became widespread in the 1960s, the first commercial liquid scintillation counter being built as early as 1953 (Taylor 1987:76,77,86; Polach 1992:199). The advantages of these techniques over the solid carbon counter was that the maximum dating range was increased as was counting precision, less sample was required, and reproducibility was improved (Taylor 1987:83).

The fourth technique, Accelerated Mass Spectrometry (AMS), was developed in the late 1970s (Taylor 1987:90). While the previous techniques measure the $^{14}C$ radioactivity of a sample through detection of beta particle emissions, AMS actually counts the number of $^{14}C$ atoms in a sample. Though only very small samples are required for this technique, it is expensive in comparison to β decay methods.

The University of Waikato Radiocarbon Dating Laboratory obtains radiocarbon dates by LS counting of benzene. For this thesis, I converted 100 marine shell and charcoal samples to benzene for LS counting. The benzene ($C_6H_6$) was produced through a series of chemical reactions in vacuum systems following procedures as set out by Hogg (1982) and Hogg *et al.* (1987). An outline of the procedure follows.

Carbon dioxide was first produced either by acid hydrolysis of carbonates (using dilute HCl) or by combustion of organic samples:

$$\mathbf{CaCO_3 \xrightarrow[HCl]{} CO_2}$$

$$C + O_2 \xrightarrow{heat} CO_2$$

During combustion of the organic samples, the gases were passed through a purification train. This included heated copper oxide (CuO) to aid in the oxidation of CO, NO, $N_2O$, $NO_2$ and tar materials, silver nitrate ($AgNO_3$), to precipitate any halogens, mercuric nitrate ($Hg(NO_3)_2$) to remove halogens still present, and potassium iodide in iodine ($KI/I_2$) and chromic acid ($K_2Cr_2O_7$) to remove nitrogen and sulphur compounds. Water vapour was removed by silica gel and cooled ethanol traps (-80°C). Carbon dioxide produced by acid hydrolysis of carbonate was cleaned only by the removal of water vapour.

The Waikato laboratory utilises 3ml (*ca.* 2.64g) of benzene for counting standard samples. This amount is generated by approximately 33 grams of carbonate, and 8 to 10 grams of charcoal or wood.

The $CO_2$ produced through hydrolysis or combustion was then slowly added into a vessel containing 17g of molten lithium (Li), producing lithium carbide ($Li_2C_2$):

$$2CO_2 + 10Li \rightarrow Li_2C_2 + 4Li_2O$$

When reaction of the $CO_2$ with the lithium was complete, the lithium carbide was heated at 800°C for 30 minutes under active vacuum to remove unreacted gases. The $Li_2C_2$ was then left to cool to room temperature. When cool, distilled water was added to the $Li_2C_2$ to produce acetylene ($C_2H_2$):

$$Li_2C_2 + 2H_2O \rightarrow C_2H_2 + 2LiOH$$

The acetylene was passed through a phosphoric acid trap to remove ammonia compounds, and a cooled ethanol trap to remove water vapour. The acetylene was then transferred to a cooled vanadium-activated silica-alumina catalyst column where it was trimerised to benzene:

$$3C_2H_2 \rightarrow C_6H_6$$

The benzene was then collected by heating the catalyst in an 100°C oven, and freezing down the benzene vapour produced under active vacuum at -65°C. Upon thawing, the benzene was then transferred into a teflon backed screw-capped vial, and placed in a freezer for storage. Benzene samples are generally stored for one month to allow the decay of radon ($^{222}Rn$) to occur if it should be present (the half-life of $^{222}Rn$ is only 3.82 days) (Gupta and Polach 1985:108-109). This is important as this isotope also emits β particles of similar energy to $^{14}C$ beta emissions, therefore interfering with their measurement and altering the apparent radiocarbon age.

After one month the benzene was thawed and then transferred into a standard Waikato Synthetic Silica 3ml counting vial (Hogg *et al.* 1991:123). The scintillator butyl-PBD ($C_{24}H_{22}N_2O$) (see Polach 1992:200-201), was then dissolved in the sample benzene at a concentration of 15g/l. After dissolution of the scintillator, the benzene sample was placed in a Wallac 1220 Quantulus spectrometer (see Lowe and Hogg 1992; Higham 1993). Benzene samples for this dissertation were counted for 3000 minutes to achieve a standard error of ± 40 years.

In deriving a radiocarbon age from an archaeological sample, we measure its count rate and compare this to modern standard and background count rates. The two international modern standards used are HOxI (Oxalic acid I) and HOxII (Oxalic acid II) (Long 1995:iii). An important secondary standard is ANU (Australian National University) sucrose (see Stuiver and Polach 1977:355; Gupta and Polach 1985:103). All three standards are employed at the Waikato laboratory (Hogg 1982:40-43; Gupta and Polach 1984:104). By international convention, the modern standard is taken as 95% of the net activity of the OxI standard. This activity is equal to the count rate of 1890 wood, corrected for decay to 1950 (Gupta and Polach 1985:103). With the new oxalic acid standard HOxII, "0.7459 times the new oxalic acid activity (when normalised to $\delta^{13}C_{PDB}$ = -25‰) equals 0.95 times the old oxalic acid activity (normalised to $\delta^{13}C_{PDB}$ = -19‰)" (Stuiver 1983:794). The year 1950 is used as 0 years BP in radiocarbon dating, as this year was close to the publication of the first radiocarbon dates by Arnold and Libby in December of 1949 (Taylor 1987:5). Later, the year 1950 had an additional use as it was after this time that many atomic tests were undertaken which affected the natural $^{14}C$ level in the carbon exchange reservoir.

All *conventional radiocarbon ages* (or CRA[9]) used in this thesis were calculated according to Stuiver and Polach (1977:355-363) and therefore include the following (Stuiver and Polach 1977:356):

"A. the use of the 5568 yr half-life (mean life 8033y),

B. the assumption of constancy of $^{14}C$ atmospheric level during the past,

C. the use of oxalic acid (direct or indirect) as a standard,

D. isotopic fractionation normalisation of all sample activities to the base of $\delta^{13}C$ = -25‰ (relative to the $^{13}C/^{12}C$ ratio PDB) [PDB being the Pee Dee Belemnite from the geological formation of this name in the United States (Higham 1993:23). Recently, the PDB nomenclature has changed to VPDB as this standard has been altered slightly (Long 1995: iii)],

E. the year 1950 is automatically the base year, with ages given in years BP (ie, present is AD 1950)."

Isotopic fractionation for each sample radiocarbon dated was determined by a VG Micromass 602C Mass Spectrometer at the University of Waikato.

Conventional radiocarbon ages are calculated using the radiocarbon decay equation:

$$t = -8033 \ln(A_{SN}/A_{ON})$$

where -8033 is the mean lifetime of $^{14}C$, $A_{SN}$ the net sample count rate, and $A_{ON}$ the net activity of 0.95 HOxI.

9 Where CRA is found in text or in tables in this study it refers to the Conventional Radiocarbon Age.

The resulting conventional radiocarbon age calculated is given in years BP and has an associated 1σ standard error, for example 600 ± 40 years BP. At the Waikato laboratory, CRAs of less than 1000 years are rounded to the nearest 10 years, and the associated error to the nearest 5 years up for errors less than 50 years (see Stuiver and Polach 1977:362). If the reservoir corrected radiocarbon age is less than 200 years BP, then by convention, the age is termed 'Modern'.

## STATISTICAL ANALYSIS AND CALIBRATION OF CONVENTIONAL RADIOCARBON AGES (CRA)

All statistical calculations in this study were undertaken using the statistical techniques of Ward and Wilson (1978) for the comparison and combining of radiocarbon ages. This method was also used by Higham (1993) for pooling radiocarbon ages from New Zealand archaeological sites. Here the statistical relationship between a set of dates is found by undertaking a *T*-test on the pooled dates. Ward and Wilson (1978:20) explain that when pooling radiocarbon ages there are two Cases for consideration. In Case 1, the radiocarbon ages come from the same object or parts of the same object; in Case 2 the ages are not known or cannot be presumed to be from the same object. In combining radiocarbon ages from a single layer in an archaeological site, we use the Case 1 consideration. To test the hypothesis that the radiocarbon ages are similar, we pool our set of dates by using the formula:

$$A_p=\left(\sum_1^n A_i/E_i^2\right)/\left(\sum_1^n 1/E_i^2\right)$$

where *Ap* is the pooled mean radiocarbon age, *Ai* the radiocarbon age in years BP, and *Ei* is the standard deviation associated with this age (Ward and Wilson 1978:19,21). To evaluate whether the pooled ages are statistically indistinguishable, we use the *T*-test:

$$T=\sum_1^n (A_i-A_p)^2/E_i^2$$

where T has a chi-square distribution of n-1. The *T*-test may be undertaken at the 1 or 2 sigma probability confidence level. In this study, all statistical comparisons are at the 2 sigma level (95%) (P = 0.05). If the ages are not statistically different then they may be combined and the variance calculated using the formula:

$$V(A_p)=\left(\sum_1^n 1/E_i^2\right)^{-1}$$

where $V(A_p)$ "is the variance of the *mean* of the group of *n* observations and *not* the variance of the group of determinations." (Ward and Wilson 1978:23).

When establishing whether multiple layers in an archaeological site were deposited at the same time, a Case 2 consideration is used. This is because each dated layer represents a single event (or object), and so all layers cannot be presumed to be contemporary. Calculating *Ap*, *T*, and V(*Ap*) in the Case 2 consideration uses the same formulae as for Case 1, but as each layer represents an event at a particular point in time expressed by its radiocarbon date, the error associated with the calibration curve (*Fi*) must be taken into account for each age, along with the counting error (*Ei*) (see Ward and Wilson 1978:21-23). Therefore, the variable $E_i^2$ in the formulae given above for Case 1, is replaced by the variable $S_i^2$, where $S_i^2 = E_i^2 + F_i^2$ (Ward and Wilson 1978:23)

All conventional radiocarbon ages in this dissertation were calibrated to years AD at 1σ using the calibration program CALIB 3.0.3 developed by Stuiver and Reimer (1993). This program provides two internationally accepted atmospheric calibrations curves for the calibration of terrestrial based samples. For New Zealand archaeological charcoal samples, the Stuiver and Becker (1993) decadal atmospheric calibration curve was used as opposed to the bidecadal, as this curve is based on samples with short lifespans and little inbuilt age, whereas the bidecadal curve is applicable to samples with longer lifespans and unknown inbuilt age (Higham 1993:25). A southern hemisphere offset of -40 $^{14}C$ years was applied to charcoal CRA before calibration, as recommended by Vogel *et al.* (1993:74).[10]

Marine shell conventional radiocarbon ages were calibrated using two different techniques to ascertain which method is more suitable for calibrating New Zealand marine shell samples (see Chapter 5). Firstly, marine shell CRAs were calibrated by subtracting the current New Zealand Reservoir Correction of 385 years (McFadgen 1978:384-390), and then calibrating the resulting age on a decadal atmospheric calibration curve (in this case using the CALIB 3.0.3 Stuiver and Becker (1993) curve). This technique has been used for many years to calibrate New Zealand marine shell CRAs. Secondly, marine shell conventional $^{14}C$ ages were calibrated using the Stuiver and Braziunas (1993) modelled marine curve (also available with the CALIB 3.0.3 program), and applying a ΔR correction of -25 ± 15 years as recommended by Higham and Hogg (1995) (see Chapter 5). This marine curve bases its calibrations on a model of diffusion and movement of $^{14}C$ through the world's oceans. It has been argued that the modelled marine curve provides a more realistic method of calibrating marine shell conventional radiocarbon dates into calendar dates, as the calibrations are based on the samples' own $^{14}C$ reservoir parameters (see Stuiver *et al.* 1986; Stuiver and Braziunas 1993).

[10] since completion of this research, CALIB 4.0 (see Stuiver and Reimer 1993) has been released which incorporates the new Stuiver *et al.* 1998a and Stuiver *et al.* 1998b calibration datasets. In addition, preliminary results on current research regarding the southern hemisphere correction suggests this correction may now be -27 $^{14}C$ years (McCormac *et al.* 1998). The revised CALIB 4.0 programme and the proposed new southern hemisphere correction do not affect the results of this research.

# CHAPTER 3

# THE SAMPLE TYPE DILEMMA IN RADIOCARBON DATING NEW ZEALAND ARCHAEOLOGY

*"It must be apparent from these various arguments that there is no systematic margin of error between radiocarbon dates on samples of different type." Anderson (1991:781)*

## INTRODUCTION

Possibly the most common method used by archaeologists today for placing an archaeological event in its prehistoric time frame is radiocarbon dating. In countries where prehistory extends back over thousands of years, the use of differing sample types for radiocarbon dating an archaeological site can affect the chronological outcome, but not to the scale at which it could here in New Zealand. With human occupation comprising about 900 years, problems encountered in using various materials for dating prehistoric events within this period may be far more significant. It is therefore vital for New Zealand prehistory that variations in radiocarbon age between differing sample types from the same stratigraphic context be investigated.

This chapter presents a historic review of past investigations and opinions on differences between radiocarbon dates from various materials, and the reliability of particular sample types, for dating in New Zealand archaeology. It will be seen that investigations of sample type have been few in number, even though opinions on the reliability of various materials have been prolific.

## INVESTIGATIONS AND OPINIONS ON SAMPLE CHOICE AND RELIABILITY IN RADIOCARBON DATING NEW ZEALAND ARCHAEOLOGY: A HISTORICAL REVIEW

### *Lockerbie (1957)*

Observations that different sample types from the same archaeological context exhibited varied radiocarbon ages extend back to the 1950s. Lockerbie (1980) began experimenting in 1957 with the suitability of charcoal, marine shell, and seal bone for dating the Pounawea archaeological site in the South Island (Figure 3.1). The site contained three cultural layers: an Upper layer having a reservoir corrected marine shell age of 290 ± 60 years BP (NZ-54); an Intermediate layer with a reservoir corrected marine shell age of 500 ± 60 years BP (NZ-57), a charcoal age of 520 ± 55 years BP (NZ-55), and a seal bone age of 550 ± 55 years BP (NZ-56); and a Lower layer with a charcoal date of 810 ± 60 years BP (NZ-58). A tree growing on top of the site was dendrochronologically dated to 224 years ago. Lockerbie (1980) noted that these radiocarbon dates supported the cultural sequence of the site, and the different sample types were all in chronological agreement.

However, in 1965, Lockerbie received amendments to his earlier 1957 marine shell ages from the Upper and Intermediate layers at Pounawea (Lockerbie 1980). The marine shell CRAs had been "corrected" by the IGNS radiocarbon laboratory according to new radiocarbon calculation methods. The recalculated marine shell age in the Intermediate layer was now older than its paired charcoal and seal dates, with the new reservoir corrected age of 660 ± 60 years BP. Lockerbie (1980) could not explain why the marine shell age would be older than the other sample types dated. This initial observation on differences between sample types by Lockerbie was never resolved, and the controversy of the chronology of Pounawea continued for 23 years.

### *Trotter (1967, 1968)*

It was not until two years after Lockerbie's 1965 observations, that other archaeologists began to investigate the sample type problem. In 1967, Trotter expressed his concern in a memo to Duff (at the time both were archaeologists at the Canterbury Museum) over age variations between charcoal, bone collagen, and reservoir corrected marine shell samples from archaeological sites in the South Island (Leach 1967) (Table 3.1). He noted, "...the age of charcoal samples is almost consistently 300-400 years earlier than collagen or shell from the same level", believing that, "In some cases this is due to the use of old or relic timber, but this cannot apply where twigs are selected." (Trotter 1967c).

In 1968, Trotter further investigated irregularities between radiocarbon ages on various materials from excavated archaeological contexts. He considered again the radiocarbon chronologies of the moa-hunting sites Tai Rua, Woolshed Flat, and Redcliffs, and the *pa* at Katiki Point, and additionally the radiocarbon dates from Moa-bone Point Cave (Table 3.2). Trotter (1968) believed that twigs and small diameter pieces of charcoal should yield accurate radiocarbon dates, but found that the charcoal CRAs from these sites were consistently several centuries older than reservoir corrected radiocarbon dates on other materials. Trotter (1968) proposed that the charcoal dated from Tai Rua, Woolshed Flat, Redcliffs, Moa-bone Point Cave and Katiki Point came from relic timber, probably driftwood, and this was the reason for the vast age differences between the marine shell and moa bone collagen radiocarbon dates, and charcoal ages. He suggested that only the marine shell and collagen dates should be accepted as reliable (Trotter 1968:86).

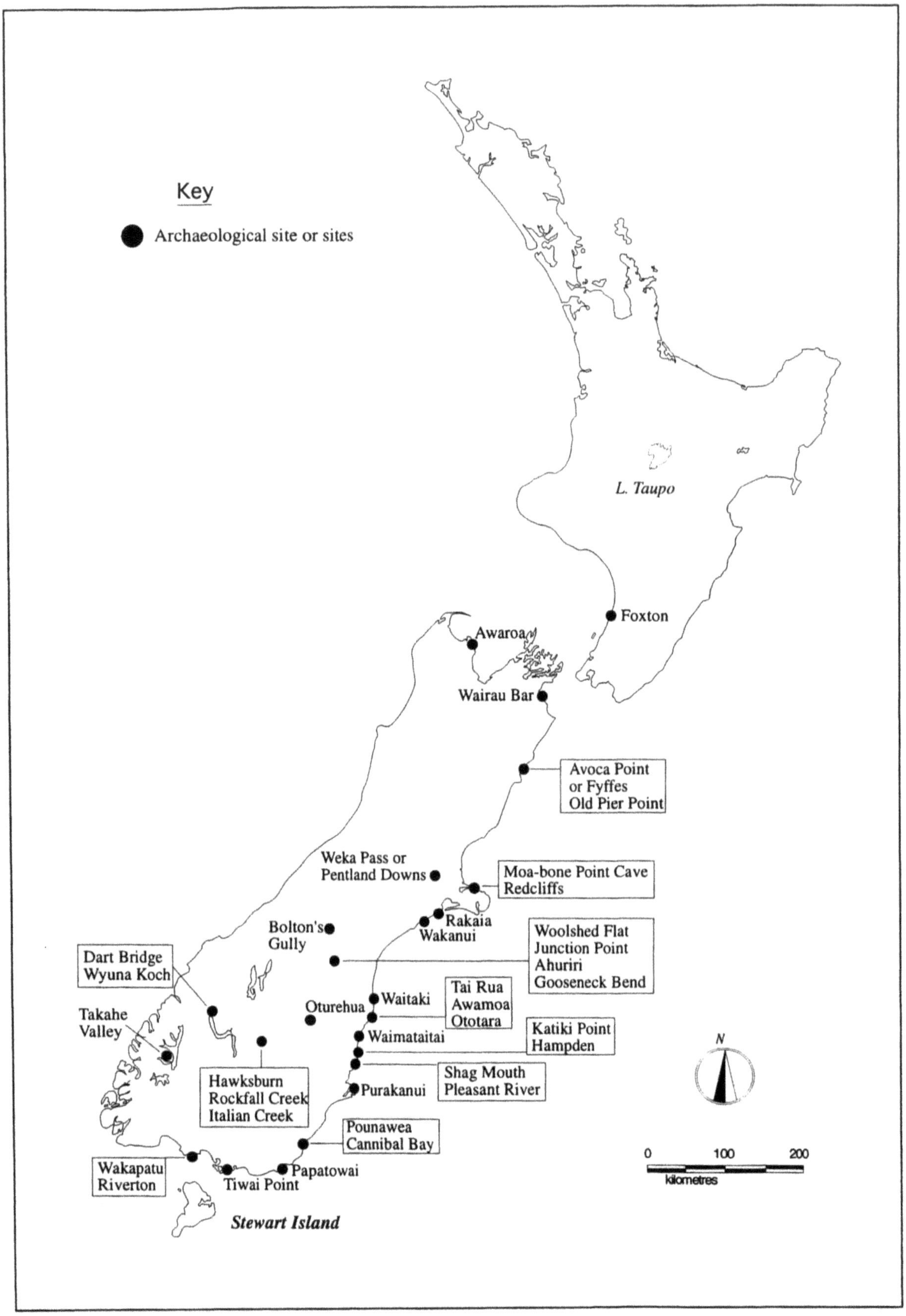


Figure 3.1. Locations of archaeological sites discussed in Chapter 3.

| Site Name | Charcoal (BP) | Marine shell (BP) | Moa bone collagen (BP) |
|---|---|---|---|
| **Tai Rua (S136/1)** | 831 ± 33 (NZ-594) | 485 ± 37 (NZ-749) | 503 ± 32 (NZ-578) |
| **Woolshed Flat (S117/3)** | 860 ± 50 (NZ-798) | - | 493 ± 70 (NZ-760) |
| **Redcliffs (S84/76)** | 1170 ± 60 (NZ-438) | 617 ± 34 (NZ-1111) | 735 ± 56 (NZ-1113) |
| **Katiki Point (S146/4)** | 880 ± 39 (NZ-696) | 211 ± 56 (NZ-697) | - |

Table 3.1. Variations in radiocarbon age on different sample types from New Zealand archaeological sites observed by Trotter (1967c) (marine shell radiocarbon dates are reservoir corrected).

| Site Name | Charcoal | Marine shell | Moa bone collagen |
|---|---|---|---|
| **Tai Rua (S136/1)** | 1119 ± 33 AD (NZ-594) | 1465 ± 32 AD (NZ-749) *(Haliotis iris)* | 1447 ± 32 AD (NZ-578)<br>1447 ± 32 AD (NZ-578)<br>1407 ± 32 AD (NZ-752) |
| **Woolshed Flat (S117/3)** | 1090 ± 32 AD (NZ-798)<br>1140 ± 50 AD (NZ-783) | - | 1457 ± 70 AD (NZ-760) |
| **Redcliffs (S84/76)** | 780 ± 65 AD (NZ-438) | 1333 ± 34 AD (NZ-1111) *(Austrovenus stutchburyi)* | 1215 ± 56 AD(NZ-1113) |
| **Moa-bone Point Cave (S84/77)** | 1163 ± 82 AD (NZ-459) | - | 1460 ± 90 AD (NZ-460) (Carbonate) |
| **Katiki Point (S146/4)** | 1070 ± 39 AD (NZ-696) | 1739 ± 56 AD (NZ-697) *(sp.?)* | - |

Table 3.2. Trotter's (1968) comparison of charcoal, marine shell and moa bone radiocarbon dates from New Zealand archaeological sites. (Trotter subtracted each reservoir corrected CRA from 1950 AD to derive years AD).

In conclusion Trotter (1968:86-87) stated:

*"...on this basis I am suspicious of the accuracy of any charcoal date not supported by dates obtained from other materials or methods...In view of the uncertainties involved there would appear to be little point in submitting archaeological charcoal for radiocarbon dating except as a last resort when no other materials - such as shell or bone collagen - are available."*

***Law (1971)***

Law (1971) considered the effect of solar cycles and nuclear tests on radiocarbon dates. He suggested that twigs, leaves and seeds should not necessarily be regarded as reliable for dating because $^{14}$C fluctuations in the biosphere every 11 years (due to solar cycles) may affect the short time span of growth. Based on this, he believed that sapling sized twigs which show a good number of years of growth (ie. slower growing), and bone collagen from large fauna are the most reliable for dating (Law 1971:65). Law noted that marine materials were immune to solar effects, but freshwater shell and moa crop material would be affected.

***Coutts (1972)***

Coutts (1972:516-531) discussed problems associated with dating various sample types by looking at 27 radiocarbon dates from 15 sites. He noted that charcoal was not always exclusively older than other contemporary materials as Trotter (1968) had observed (Coutts 1972:524). Coutts observed that one third of his archaeological sites with charcoal/marine shell pairs had a reservoir corrected marine shell radiocarbon age that was older than the contemporary charcoal radiocarbon age (Coutts 1972: 527). The remaining charcoal radiocarbon dates from other sites showed great variability when multiple charcoal ages from the same layer were compared. Coutts suggested that chronological problems associated with radiocarbon dating archaeological sites may be alleviated by multiple dating and more research on the reliability of certain materials (Coutts 1972:524).

Coutts also suggested that investigations of alternative dating methods appropriate to archaeology may be useful (Coutts 1972:524).

### *Rafter, Jansen, Lockerbie and Trotter (1972)*

Rafter, Jansen, Lockerbie, and Trotter (1972) considered diverse opinions and observations regarding the reliability of various sample types for dating. For marine shell, Rafter *et al.* (1972) noted that Taylor and Berger (1967) stressed the necessity of studying the marine environment from which marine shells used for dating archaeological contexts were derived, and Mangerud (1972) had emphasised that shells are reliable if samples were selected carefully. They also observed that Valastro and Davis's (1970) comparison of 12 pairs of charcoal and reservoir corrected shell ages from archaeological sites in Texas, showed that shell was older than charcoal by 1.36 sigma for all pairs except two, but that Haynes *et al.* (1971) reported little difference between two reservoir corrected shell and charcoal ages from his study (Rafter *et al.* 1972:627-633)

Rafter *et al.* (1972) then assessed the suitability of moa bone for radiocarbon dating. They compared carbon isotope ratios and dates of moa bone carbonate from four archaeological sites, with a contemporary bone standard. Rafter *et al.* (1972) found that isotopic fractionation in bone carbonate could introduce significant radiocarbon dating errors for archaeological samples, and that bone carbonate radiocarbon ages from some of the archaeological samples dated were affected by incorporation of bomb carbon. They concluded from their study that moa bone carbonate was not reliable for dating, but that moa bone collagen could be, as illustrated by close agreement between moa bone collagen dates and reservoir corrected shell ages from contemporary archaeological deposits (Rafter *et al.* 1972:638).

Rafter *et al.* (1972) considered a variety of opinions on the reliability of bone for radiocarbon dating. For example, Tamers and Pearson (1965) observed dates on bone carbonate were unreliable due to contamination, but Haynes (1968) and Haynes and Long (1969) stated that the bone apatite $CO_2$ dating method gave more accurate results than any other bone fraction. In dating the collagen fraction of bone, Rafter *et al.* (1972:640-643) noted Sellstedt *et al.* (1966) who believed bone collagen provided reliable radiocarbon dates (which should be more accurate than wood); Haynes (1967) who believed only collagen was reliable for dating; and finally Barker *et al.* (1971) who reported close agreement between charcoal and collagen dates when all the humic acids were extracted. Rafter *et al.* (1972:641-642) also found, however, that many studies could not decide on the reliability of bone at all, for example Barker (1967) who could make no firm decision on the suitability of bone for dating; Geyh and Guhlich's (1970) research which indicated that both collagen and apatite should be dated and compared for a more reliable result, but these dates should not be completely trusted; and Polach (1971) who dated six bone samples from one archaeological site to investigate the reliability of dating bone carbonate, apatite, and residual organic carbon, and found no agreement between dates on these fractions (Rafter *et al.* 1972: 642).

Rafter *et al.* (1972) then investigated differences in radiocarbon age between sample types from New Zealand archaeological sites (Figure 3.2). Forty four radiocarbon dates from 11 archaeological sites and natural swamp deposits of moa bone were compared, the plant material being contemporary with the bone collagen and shell (Rafter *et al.* 1972:631). From the analysis, radiocarbon dates from bone carbonate were found (as expected) to be unreliable. Plant remains were consistently older than that of shell or bone collagen, but the shell and bone collagen radiocarbon ages were in close agreement. Explanations suggested for the age differences between the plant remains and associated shell and bone, were the use of relic logs by the sites' inhabitants (such as at Katiki Point (totara)), or that the charcoal used for dating possibly came from the outer and inner rings of ancient trees (as proposed for the Redcliffs and Moa-bone Point Cave dates).

Rafter *et al.* (1972) concluded from their comparison of sample types, that plant remains usually give radiocarbon dates several hundred years older than reservoir corrected shell or bone collagen dates. They believed that marine shell and bone collagen were more reliable for dating New Zealand archaeological sites (Rafter *et al.* 1972:644).

### *McCulloch (1973)*

In 1973, McCulloch (1973:130-131) proposed a protocol for obtaining reliable samples for radiocarbon dating. Of the four points in her protocol, the two most important were that the material to be dated must be accurately identified, and that the time of death of the material be contemporary with time of use at the site. She illustrated these two points using two radiocarbon dates from the Weka Pass site in North Canterbury; one from marine shell (*Amphidesma australis*) of 910 ± 132 years BP (NZ-1534), and the other from charcoal (*Podocarpus totara-hallii* ) of 1315 ± 80 years BP (NZ-1535). McCulloch (1973:131-132) explained that because charcoal dates were usually several hundred years older than marine shell, and if this was allowed for in this instance, then the charcoal age was in agreement with the shell age. McCulloch suggested that when looking for early sites, "I am convinced that the best and most authentic dates in this direction are to be got from small isolated sites showing only one period of occupation - perhaps one isolated visit - rather than from the large and obviously well-established sites, even at their lowest levels." (McCulloch 1973: 131).

### *Grant-Taylor (1974)*

Grant-Taylor (1974) presented information regarding the reporting and collection of samples for radiocarbon dating. He warned archaeologists against collecting charcoal for dating because of inbuilt age, as illustrated by Rafter *et al.* (1972). Grant-Taylor (1974: 159) believed that the best material for dating was short-lived materials, such as shell, bone or flax cordage, which was directly involved with the prehistoric event being dated. He also noted that if dating charcoal or wood, only twigs should be used, and when using shell, that aragonite should be tested for recrystallisation by XRD.

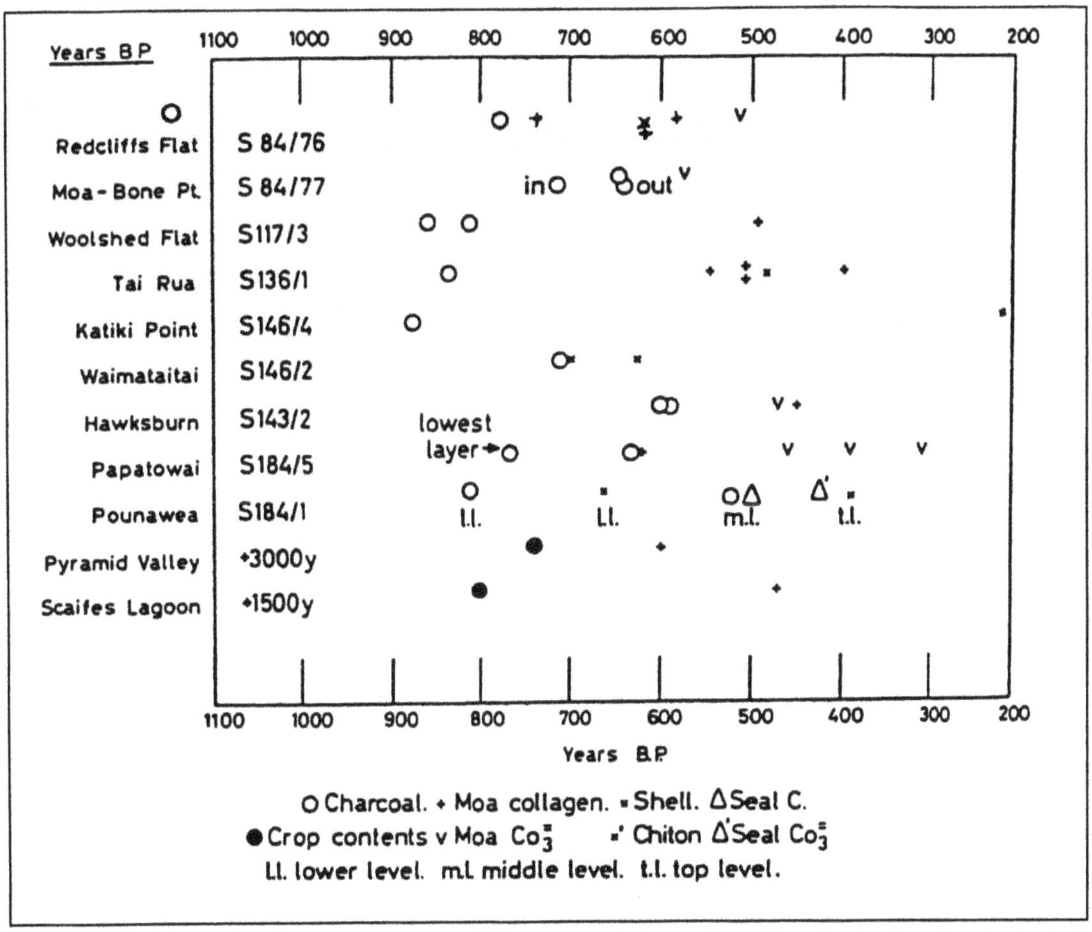


Figure 3.2. Comparison of radiocarbon ages on different sample types from New Zealand archaeological sites by Rafter *et al.* (1972).

Grant-Taylor (1974: 159) explained that shellfish such as pectens, mussels, and oysters should not be used for dating as recrystallisation cannot be detected, and *Amphibola* should be avoided as it ingests old carbon from the environment. He also suggested that:

- when dating human and moa bone, carbonate should not be dated, and that collagen was preferred;
- that soils were notorious for dating;
- fish could be a likely source for dating archaeological sites that had good stratigraphy or distinguishable artefacts, though species identification was important;
- and finally that whale bone was not a useful constituent for archaeological dating, but seal bone had been noted as producing dates which agreed closely with other materials believed to be reliable (Grant-Taylor 1974: 159-160).

A guideline presented by Grant-Taylor (1974:161-162) contained 13 steps for effective dating of archaeological sites. In summary, he believed that: 1) samples should not come from isolated single deposits, but from multi-layered sites where samples can be collected from all layers (cf. McCulloch 1973:131); 2) charcoal collections should be from twigs; 3) shells appeared to be very good for dating, but in certain areas caution should be taken, such as the Cook Strait to Kaikoura coastal region where oceanic upwelling may occur; 4) fish bone may be used, especially vertebrae, if they can be identified and aged; 5) moa bone was good for dating, and may directly date moa-hunting; 6) identified small bird bone can be used, but land and sea birds should be kept separate; and 7) a sample should be carefully associated with the event to be dated, and a stratigraphic column provided so that the position of each sample may be marked.

### *McCulloch and Trotter (1975)*

McCulloch and Trotter's (1975) study on the chronological distribution of moa-hunting sites in the South Island highlighted the sample type problem. Of the 180 radiocarbon dates available from South Island archaeological sites, 88 were used in their analysis. Dates and sites considered dubious by the authors were withdrawn based on a protocol which considered the unreliability of bone carbonate, unreliable provenance of samples to moa-hunting sites, dates from relic timber, and lack of firm evidence for moa-hunting associated with the date. The radiocarbon dates were then analysed by sample type, providing 20 moa bone collagen, 18 marine shell, and 44 charcoal radiocarbon dates, with 6 radiocarbon ages coming from other materials. This analysis produced perhaps one of the most well known illustrations on sample type differences in New Zealand archaeology (Figure 3.3).

McCulloch and Trotter (1975:5) viewed moa bone as a reliable indicator of initial occupation of a site, and believed

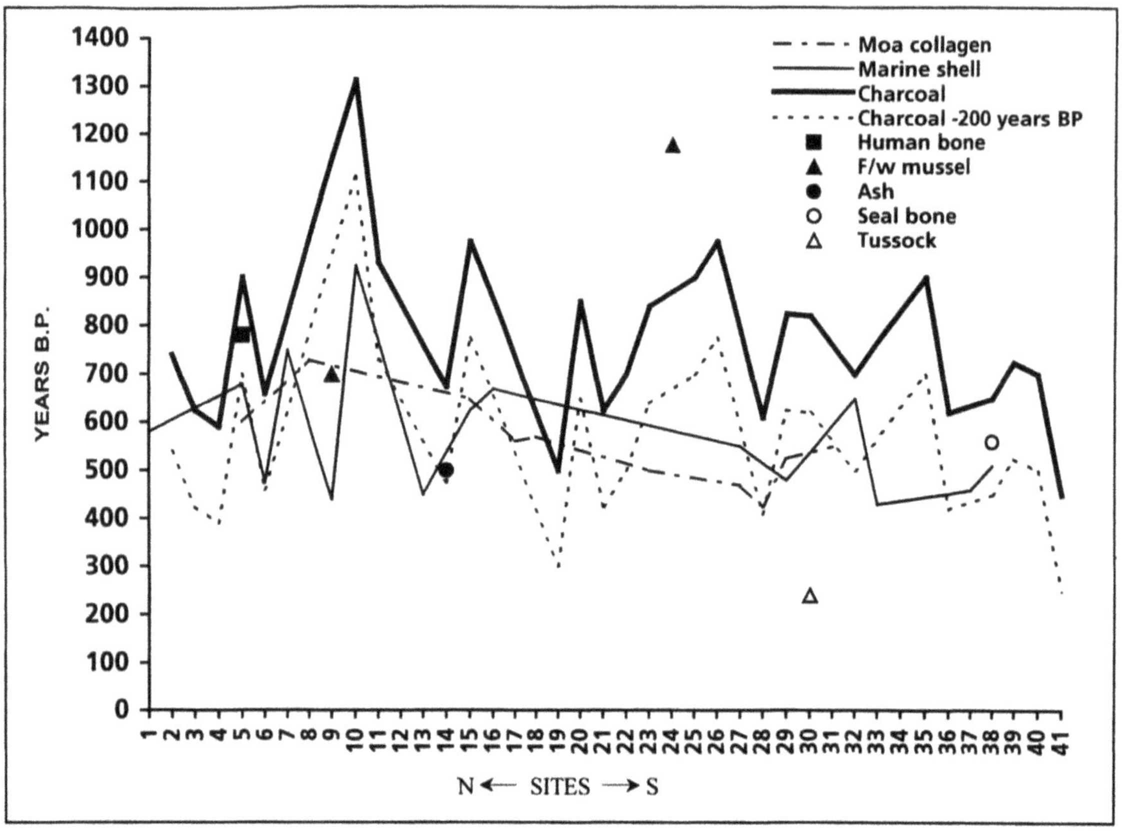


Figure 3.3. Comparison of radiocarbon ages on charcoal, marine shell and moa collagen from New Zealand archaeological sites by McCulloch and Trotter (1975) (black and white reproduction of original colour graph).

that it was not prone to contamination from the atmosphere or surrounding soil, or at risk of compositional change. The only disadvantage of dating moa bone was its availability from archaeological deposits.

They considered marine shell to be problematic due to geographical variations in reservoir corrections, and potential for recrystallisation. Charcoal and wood were viewed as possessing many problems in terms of sample history, such as the use of relic timbers by a site's inhabitants. McCulloch and Trotter suggested that charcoal and wood should be identified before dating to isolate short-lived material, but that charcoal was at risk of contamination by humic materials (1975:9).

From their analysis of sites where charcoal had been dated alongside either marine shell or moa bone collagen (Figure 3.3), McCulloch and Trotter (1975:11-13) observed that charcoal dates were consistently older (by on average 200 years) than dates on these materials. They concluded that moa bone collagen was the most reliable dating material, followed by marine shell. They also suggested that charcoal was unreliable (McCulloch and Trotter 1975:14).

***Law (1981)***

Law (1981) provided guidance for researchers in selecting materials for radiocarbon dating. He noted that marine shell was excellent for dating, giving consistent and reproducible ages (Law 1981:233). Also, that contamination of marine shell samples could be investigated using x-ray methods, but *Amphibola crenata* should never be dated (Law 1981:233). Law recommended that only identified charcoal from short-lived species or young wood be utilised, but warned that contamination was possible from humic acids and the effective removal of these was disputed (Law 1981:233). Additionally, he noted that identified charcoal samples were often made up of three or more species, and unidentified charcoal often contained wood up to 200 years old (Law 1981:233-234). Law (1981:234) believed that marine fish bone should be excellent for dating as long as pelagic fish were avoided as these species can feed outside of New Zealand waters. He also suggested that human bone should not be used because of the possibly high marine diet. Moa bone was seen by Law as acceptable (as well as bone from animals with a sole diet component), but he noted that doubts on the contamination of collagen were unresolved. For marine mammals and birds, Law (1981:235) proposed that animals which derive their food from tropical or Antarctic seas should not be dated.

***Anderson (1982)***

In discussing the chronology of economic patterns in Archaic southern New Zealand, Anderson (1982:47) noted (see Figure 3.4):

*"The chronology of the southern Archaic is at present in a state of some confusion because radiocarbon dating in this region sometimes produces estimates up to 300 years apart for similar layers depending on the material processed (Trotter 1968, McCulloch and Trotter 1975)....To what extent this discrepancy is a real one is unknown since few sites have been carefully dated by a range of materials and many of the earlier dates were obtained on*

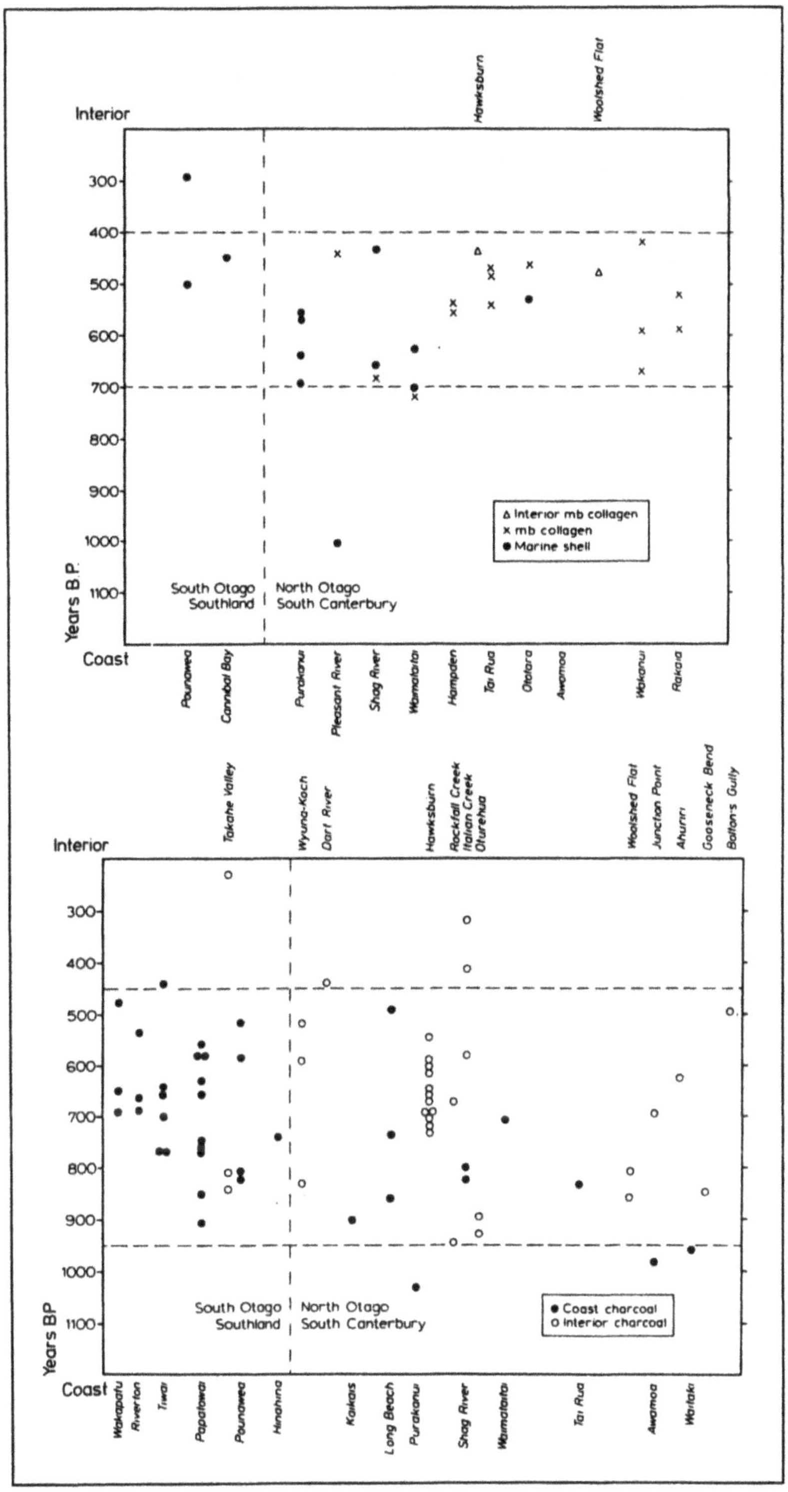


Figure 3.4. Comparison of sample types used to date New Zealand southern Archaic archaeological sites by Anderson (1982).

*charcoal from wood of unknown age."*

Anderson (1982:47) made two significant observations in his discussion of the charcoal chronology of the southern Archaic. Firstly, he noted that even though charcoal radiocarbon ages from some Archaic sites had been derived from short-lived material, there was no reason for rejection of the charcoal chronology which extended back to 800-900 years BP until more difficult problems with the dating of charcoal were found (Anderson referring to McCulloch and Trotter's (1975:8) view that inbuilt age greatly affected charcoal radiocarbon chronologies). Secondly, Anderson suggested that if there was a materials-dependent difference in obtaining reliable radiocarbon ages, then this was not consistent, because some sites showed chronological agreement between materials dated, and others did not (Anderson 1982:47). Based on these two points, Anderson

(1982:47) believed that charcoal radiocarbon dates for the southern Archaic were acceptable.

### *McFadgen (1982)*

McFadgen (1982) identified few problems in radiocarbon dating archaeological sites using marine shell. But he did suggest that because dates on marine shell would only be as accurate as dates on charcoal or wood, shell dates with standard errors less than 50 years, should have their errors increased to 50 years (McFadgen 1982:384). From his recalculation of the New Zealand marine shell standard of 330 years (McFadgen 1978) (see Chapter 5), McFadgen proposed that as there had been little change in the radiocarbon activity of the mixed layer ocean in the past 1000 years, the subtraction of a further -55 years from the current marine shell standard (now making the standard 385 years) for shells less than 1000 years old, should provide good calendar dates (McFadgen 1982:384). McFadgen (1982:384) next considered paired charcoal and wood radiocarbon dates from a number of archaeological sites to illustrate that age differences between contemporary charcoal and wood dates was due to inbuilt age.

### *Law (1984)*

In 1984, Law investigated the reliability of marine shell for radiocarbon dating New Zealand archaeological sites. He observed that even though by 1984, 191 dates existed on shell from archaeological sites, this material had not been considered acceptable by archaeologists. Law suggested that this was because marine shell had not been seen as a conventional material to use, that it had no calibration curve to correct ages with, and that the different marine shell corrections employed in differing areas of the world were not understood by archaeologists (Law 1984:1). Law considered the current marine shell standard for New Zealand (336 years), and noted that from the seven modern shells dated, no regional differences in reservoir ages could be seen (Law 1984:3). He noted that *Chione stutchburyi, Paphies australe, Paphies subtriangulatum, Paphies ventricosum, Haliotis iris, Turbo smaragda,* and *Protothaca crassicosta* were preferable marine shell species for dating, and that *Amphibola crenata* should be avoided. Law illustrated that marine shell dates from New Zealand archaeological sites with multiple cultural layers did show site stratigraphic agreement, and that marine shell was reproducible. He then compared age differences between moa bone collagen, reservoir corrected marine shell, and charcoal dates from the same archaeological context at a number of archaeological sites (Law 1984:5-6). By noting the difference in years between the sample types for each site and averaging these, Law found that moa bone collagen ages agreed closely with reservoir corrected shell dates, with a mean difference of 0 ± 25 years. The differences between charcoal and reservoir corrected shell were greater (up to 553 years), even though the mean difference was only 83 ± 36 years.

Law concluded from his study that due to problems such as the Suess effect, charcoal and collagen ages of less than 450 BP were useless (Law 1984:6-7), but shell was useful up to 250 years or less. Also, that the often remarked difference in radiocarbon age between charcoal and other sample types of 200 years, was on average only 83 years, and there were good reasons to prefer shell than charcoal for dating the prehistory of New Zealand (Law 1984:7).

### *Trotter and McCulloch (1984)*

Trotter and McCulloch (1984:718) made a clear distinction with regards to the accuracy of charcoal, marine shell and moa bone collagen for dating moa-hunting sites. Trotter and McCulloch believed that although charcoal was the most common material used for radiocarbon dating these sites, the death of the wood rarely dated the use of a fire (1984:718). Two possible reasons for this were suggested. Firstly, that wood used for fuel probably originated from trees long dead. Secondly, that even if living wood was used, its inner core would be of different age than that of its outer wood or associated branches. They also suggested that charcoal was highly susceptible to contamination from soil constituents, as experienced at Redcliffs and Tai Rua (Trotter and McCulloch 1984:718). They believed that moa bone collagen was a better material for dating moa-hunting sites as it appeared to resist contamination from the ground or the atmosphere, and had shown very consistent results (Trotter and McCulloch 1984:718). Coastal marine shell was seen as producing similar ages to moa bone collagen, but lacked reproducibility (Trotter and McCulloch 1984:718).

Trotter and McCulloch (1984) observed that moa bone collagen and charcoal dates from the same site, on average, diverged in age by as much as 300 years. Based on this observation and those made above, they proposed, "In discussing the date of moa extinction and the part played by man in that extinction, we would prefer to consider only samples of moa bone collagen" (Trotter and McCulloch 1984:718). But due to the lack of collagen dates from moa-hunter sites, Trotter and McCulloch also employed marine shell ages for their analysis, excluding dates from *Amphibola* and shell samples noted by the radiocarbon laboratory to have been affected by recrystallisation. At the time of their analysis, 27 moa bone collagen dates and 22 marine shell ages were available from moa-hunter sites through-out New Zealand, but only a handful of dates came from the North Island (Trotter and McCulloch 1984: 718).

Trotter and McCulloch (1984:720) explained that although moa bone collagen and marine shell dates were believed to be reliable, discrepancies did occur which conflicted with the archaeological evidence from some sites. They noted that problems with the radiocarbon chronology at Wairau Bar and Papatowai, were probably due to the samples being stored for a long period of time before being used for dating (Wairau Bar), and samples being collected by various people over a period of time from different parts of the site (Papatowai). Dates from Papatowai were in general considered dubious as they were mainly taken from charcoal and moa bone carbonate, and there was disagreement between a charcoal and reservoir corrected shell date from the same level (Trotter and McCulloch 1984:720). They noted that only one moa bone collagen date came from the North Island, and some of the few sites with shell dates contained stratigraphic anomalies, such as the Foxton site (Trotter and McCulloch 1984:722).

### *Anderson (1984)*

Anderson discussed the radiocarbon chronology of moa extinction in Southern New Zealand. He observed:

*"Many dates are unpublished, and exact details of their provenance are unknown. The same may be said of too many published dates. Worse, there is frequently a discrepancy of up to 300 years between radiocarbon estimates from charcoal samples and those from other materials such as bone collagen and marine shell. The difference is inconsistent, and the cause is yet unclear." (Anderson 1984:731-733).*

Anderson (1984:733) noted that while charcoal dates extended as far back as 1000 years BP, those on collagen and reservoir corrected marine shell fell around 800 years BP, and that for some sites, dates on charcoal, collagen, and marine shell samples from the same context were in chronological agreement (Anderson 1984:733). Anderson preferred charcoal for dating the moa-hunting period, because recent dates on this material were speciated and processed for soil contamination. He also believed that such dates were the most accurate when reconstructing changes in vegetation for a location through time (Anderson 1984:733).

### *Caughley (1988)*

Caughley (1988) considered the radiocarbon chronologies from moa-hunting sites in the North and South Islands in an attempt to trace the colonisation pattern of New Zealand. His study included a critique of the use of various sample types for dating archaeological sites, assessing what material would be best for dating the movement of the early Polynesians from the area of their first landfall. He believed that contamination and false association (the death of the sample being different from the event being dated) were the major sources of error in dating a short prehistory (Caughley 1988:246). False association took the forms of Redeposition; Inclusion; Mixing in place; and Contemporary deposition of samples, but different radiocarbon ages. Caughley attributed "contemporaneous deposition" as the main cause of large age variations between charcoal, reservoir corrected marine shell and collagen radiocarbon dates. Problems associated with the various sample types identified were inbuilt age for charcoal and wood, and young ages for bone collagen (Caughley 1988:247).

Caughley presented radiocarbon data showing all moa-hunting sites with both marine shell and moa bone collagen dates (see Caughley 1988 Table 4). In an analysis of this data, he found that there was little difference in age between these sample types from within the sites, but that dates from marine shell and moa bone collagen varied significantly from associated charcoal ages. This, he noted, had also been observed by Trotter (1968) and Trotter and McCulloch (1973) (Caughley 1988:249).

Caughley (1988) compared Anderson's (1983) list of charcoal dates from moa-hunting sites with his own list of marine shell and moa bone collagen ages (35 dates from 10 sites which included 16 charcoal ages, and 19 marine shell and moa bone collagen dates). He found the mean of one sample type from a site and compared this with the mean of the others. The greatest age difference found between the reservoir corrected ages was 264 ± 37 years for charcoal. Next, he pooled the marine shell and moa bone collagen dates from the sites to produce a mean date on those materials, and comparing this with the pooled mean of the charcoal dates, showed charcoal to be 234 ± 44 years older than the other materials. Finally, Caughley took the two sets of dates and performed a regression analysis on the distances they were from Kaikoura (Caughley 1988:250). This method again showed charcoal was the older material by 241 ± 42 years. Caughley (1988:250) believed this analysis showed that charcoal was on average 240 years older than dates on moa bone collagen or marine shell. Caughley (1988:250) did note, however, that the difference of 240 years may be smaller as some charcoal dates were more recent and may have been screened for long-lived species.

In conclusion he stated:

*"Whether charcoal dates are more accurate than collagen and shell dates cannot presently be determined. What is not in doubt is that collagen dates and marine shell dates are consistent with each other, and jointly or separately provide a more precise relative chronology than does charcoal." (Caughley 1988:250).*

Caughley believed that moa bone had an advantage over charcoal in that it actually dated an event, the death of a moa, and it was only this one event that had to be determined.

### *Anderson (1989)*

Anderson (1989:172-173) addressed sample type differences in radiocarbon dates from moa-hunter sites. He stated that:

*"One of the long-standing uncertainties concerning archaeological chronology in New Zealand is whether there is a systematic difference in the radiocarbon results obtained on charcoal samples as opposed to those obtained on bone collagen and marine shell." (Anderson 1989:172).*

Anderson (1989:172) noted Trotter (1968) and Trotter and McCulloch's (1973) 300 year difference between charcoal and marine shell and collagen radiocarbon dates, Caughley's (1988) 240 year mean difference, and Law's (1984) average difference of 83 years. But he also noted that at some sites, such as Waimataitai, there was no great divergence in age between sample types. Anderson (1989) averaged the radiocarbon ages for each sample type from his moa-hunting radiocarbon chronology, and illustrated that a mean difference of 26 years between marine shell and moa collagen, and 77 years between marine shell and charcoal existed. From his analysis, he concluded that the difference in age between moa bone collagen or marine shell and charcoal dates was about 100 years (Anderson 1989:172). Anderson recognised that it was difficult to ascertain why there was a difference between radiocarbon ages on various sample types, and which material provided a consistent accurate chronology. He observed that one of the problems was a lack of moa bone collagen or marine shell dates from the northern North Island, southern South Island, or inland sites (Anderson 1989:173). Other problems recognised by Anderson was that wood was not identified until 1976, that collagen dates may be too young, and estuarine shell used commonly for marine shell samples, may have similar problems to fresh water shell (Anderson 1989:173). Anderson concluded that, "Until more is known about these problems...it would be prudent to leave the matter of preference open and treat sample types as separate groups." (1989:173).

### *Anderson and McGovern-Wilson (1990)*

Anderson and McGovern-Wilson (1990:43-45) investigated the use of differing sample types for radiocarbon dating moa-hunting sites, while critiquing Caughley's (1988) analysis of moa-hunting radiocarbon dates.

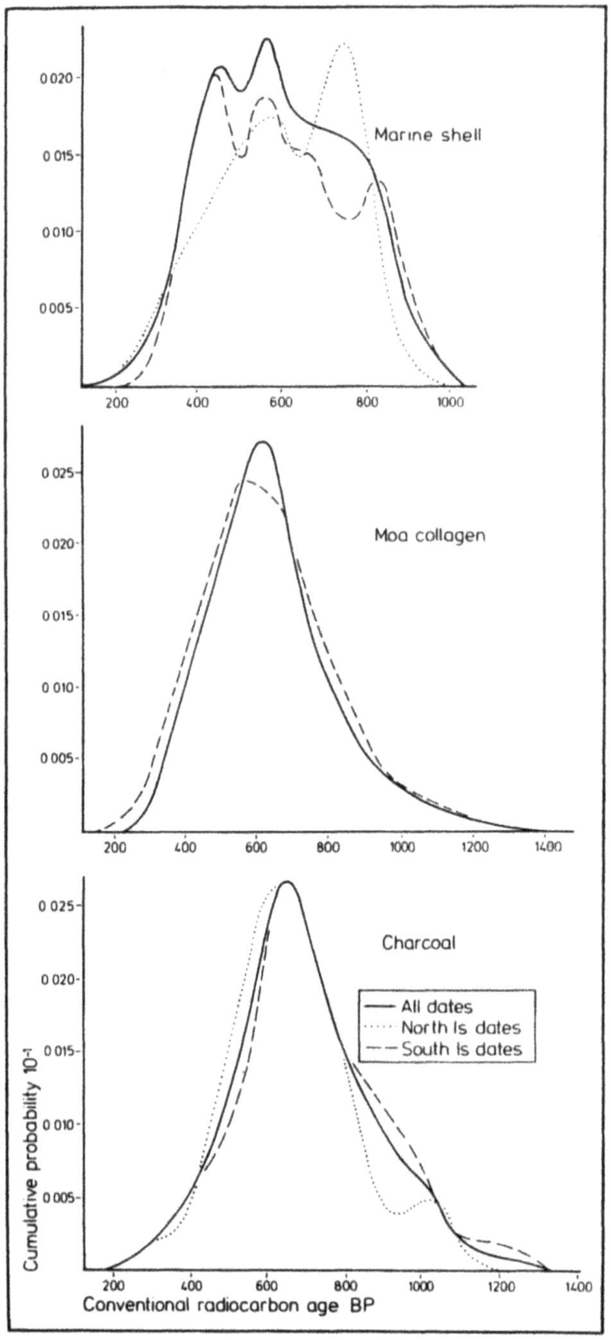


Figure 3.5. Anderson's (1989) cumulative probability curves of radiocarbon dates on differing sample types from moa-hunting sites.

Anderson and McGovern-Wilson (1990:43) observed that although Caughley believed one sample type could not be preferred over another in reflecting the chronology of New Zealand prehistory, Caughley then set out to illustrate there were distinctive age differences between charcoal, marine shell and moa bone collagen radiocarbon dates. They noted that Anderson's (1989) (see above) analysis found differences in age between charcoal dates and dates on marine shell and moa bone, were significantly less than the 240 years proposed by Caughley (Anderson and McGovern-Wilson 1990:43).

Anderson and McGovern-Wilson (1990:44) also disagreed with Caughley's observation that collagen dates were better than charcoal for dating moa-hunting sites as this material actually dated the cultural event. They explained that charcoal did have problems with inbuilt age, but since 1976 the IGNS had routinely screened charcoal samples because of this problem. They noted that moa bone could also be prone to a type of inbuilt age as the bone of birds long dead may be brought to the site for processing, as argued by Millener (1981), Prickett (1987), Nichol (1988) and McFadgen (1987). Anderson and McGovern-Wilson (1990:44) believed that this occurrence would not be of great concern for South Island sites, but was still potentially a problem as bone can survive for long periods of time in dry regions. They suggested that marine shells could also be contaminated to give older ages as observed by McFadgen

(1987) at Avoca Point, Kaikoura (Anderson and McGovern-Wilson 1990:44).

Anderson and McGovern-Wilson (1990:44) challenged Caughley's (1988:249) assertion that marine shell and collagen dates both produced very similar ages, by considering Anderson's (1989) Cumulative Probability Curves (see Figure 3.5). These curves compared charcoal, moa bone collagen and reservoir corrected marine shell radiocarbon dates from moa-hunting sites. The curves showed charcoal dates normally distributed similarly in time as moa bone collagen dates, whereas the marine shell age curve was varied. Anderson and McGovern-Wilson (1990:44) suggested this illustrated that charcoal was not "the odd man out" as observed by Caughley, and that there were many other reasons for challenging the use of various sample types, such as changes in counting standards, processing methods, or dating technique etc (Anderson and McGovern-Wilson 1990:45). Anderson and McGovern-Wilson (1990:45) concluded:

*"Nevertheless it is not our intention to dispute the validity of dates on any particular sample type; rather it is to emphasise that general uncertainty, for one reason or another, about all archaeological radiocarbon dates...renders invalid the choice of one series over another as the data base on which to test a broad chronological hypothesis."*

### *Anderson (1991)*

Anderson (1991:768) discussed the acceptability of particular sample types for radiocarbon dating in his paper on the first colonisation date of New Zealand. He noted that sets of dates from the same archaeological context show charcoal to be consistently older than marine shell and bone collagen by several hundred years, as observed by Trotter (1968), Rafter *et al.* (1972), McCulloch and Trotter (1975), Law (1984), Caughley (1988), Anderson (1989a), and Anderson and McGovern-Wilson (1990) (Anderson 1991:768). Anderson (1991:777-782) therefore addressed problems associated with the various sample types used in radiocarbon dating New Zealand prehistory, to try to determine which sample type was reliable.

Anderson suggested that there were two areas of concern in dating marine shell. The first was that most samples used for dating were estuarine species such as cockles (*Austrovenus stutchburyi*), which may have come from catchments associated with calcium carbonate of infinite radiocarbon age (Anderson 1991:777). This was shown by McFadgen's (1972, 1978) studies on differences between the radiocarbon ages of cockle and tuatua (*Paphies subtriangulatum*) from the Foxton archaeological site (Figure 3.6). The second cause of concern was the use of species which otherwise may be reliable, such as sub-tidal rocky shore shellfish, but which came from areas of deep oceanic upwelling associated with limestone strata, such as at Avoca Point, Kaikoura (Anderson 1991:777).

Anderson (1991:777-779) next considered problems associated with dating moa bone collagen. The first source of potential error was the amount of possible fractionation which could have affected the samples. He noted that although $\delta^{13}C$ levels could be determined for bone samples, determining what affect variations in fractionation meant in dating, was problematic. These could reflect anything from variations in moa diet (which should be mainly a C3 plant diet), inadequate sample preparation, or diagenetic attack from bacteria which introduce new organic compounds to the collagen while in the ground (Anderson 1991:777). Anderson observed that tests of sample integrity for bone such as "comparative dating of the amino acid fractions in the collagen, examination of carbon:nitrogen ratios (the normal range is 2:9-3:6) and the measurement of fluorine levels" had not been undertaken on a routine basis by New Zealand laboratories (Anderson 1991:779). Factors contributing to inconsistent results in his view included variable collagen concentrations and changing treatments to extract collagen (using HCl, phosphoric acid, and now 0.6N HCl) (Anderson 1991:779). Mineralisation causing young carbon to become incorporated from the surrounding environment into the bone collagen was another source of difficulty identified by Anderson (1991:779). Anderson (1991:779) also suspected that there were moa bone collagen dates on bone not associated with butchering activities, but from natural sub-fossil deposits. Anderson (1991:779) believed that because of these potential problems, radiocarbon dates on moa bone collagen could not be relied upon (Anderson 1991:779).

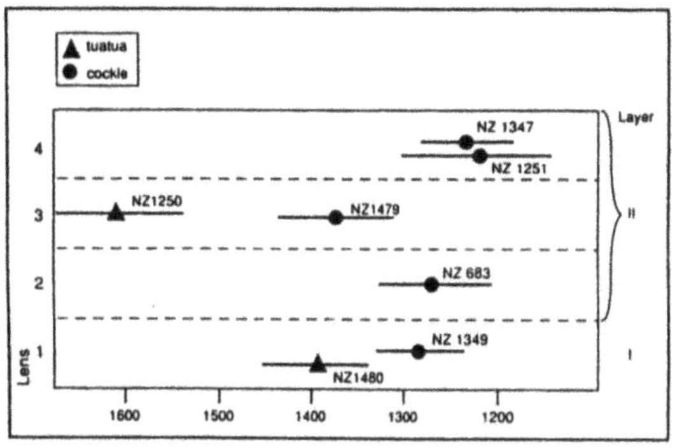


Figure 3.6. Problems in radiocarbon dating shellfish of differing species from the Foxton archaeological site (radiocarbon dates have been calibrated) (from Anderson 1991:779).

Anderson's concerns with dating charcoal came from three major areas - inbuilt age; the identification of tree species for dating; and the careful use of twigs. He emphasised that all charcoal should be identified and carefully selected prior to dating (1991:779-781; see Chapter 4). From his presentation of problems associated with the dating of marine shell, moa bone collagen, and charcoal, Anderson stated, "these various arguments [show] that there is no systematic margin of error between radiocarbon dates on samples of different type" and that sources of error can be due to local conditions (Anderson 1991:781). He concluded, "Since we cannot be sure which dates exhibit significant errors, or to what extent, there is no point in dwelling on the marine shell and bone collagen chronologies (Table 1, Figure 6). They are broadly similar to the charcoal chronology but disclose great variability." (Anderson 1991:781-782).

### *Fankhauser (1992)*

Fankhauser (1992) dated Umu-ti in South Canterbury using Thermoluminescence (TL) and radiocarbon dating to determine the settlement time of the South Island. Thirty four radiocarbon dates were obtained on identified charcoal (mainly *Podocarp* sp.) from these prehistoric structures (Fankhauser 1992:29-30). Fankhauser (1992:37) believed the radiocarbon dates indicated that South Canterbury and Southern New Zealand were colonised by *ca.* 1275 AD, but that inbuilt age had caused large variations in the charcoal ages derived from the Umu-ti. Considering alternatives to using charcoal for dating because of inbuilt age, he asked "Is there dating evidence other than radiocarbon charcoal dates with which to assess the prehistoric settlement of Canterbury and the Southern South Island?" (Fankhauser 1992: 34). In answering this question, Fankhauser excluded charcoal for archaeological dating, noting that past experience had shown it to be unreliable for dating sites, as shown by McCulloch and Trotter (1975), McFadgen (1982), and Law (1984). These studies showed that twigs can be risky to use, and even with identified wood from short-lived species, there was still an unknown inbuilt age (Fankhauser 1992:34).

Fankhauser (1992:35) then considered the use of marine shell and moa bone collagen for dating southern moa-hunting sites (using date lists from these sites such as that of Anderson and McGovern-Wilson (1990)), and observed that even though these materials appeared to be more reliable than charcoal, problems had been noted with their use. He also noted that most marine shell ages being less than 650 cal years BP may be an issue of concern to prehistorians (Fankhauser 1992:35).

In the light of his charcoal radiocarbon dates, Fankhauser advocated the use of TL for dating New Zealand archaeological sites. He believed that TL dating could provide more accurate results than radiocarbon dating, especially when attempting to radiocarbon date charcoal because of the problem of inbuilt age (Fankhauser 1992:36). Fankhauser proposed that TL would seem to be an important and sensible alternative to the dating of charcoal, as greywacke oven stones were common in archaeological sites. He concluded that although marine shell may be more reliable for radiocarbon dating than charcoal, where marine shell is lacking from a site, TL dating may provide reliable dates, and so aid in resolving the prehistoric sequence of New Zealand (Fankhauser 1992:37).

### *Higham (1993)*

Higham's (1993) research on radiocarbon dating New Zealand prehistory found few problems in the dating of charcoal and marine shell from archaeological contexts. This was the first detailed study which looked at the differences between identified charcoal and marine shell ages, and considered three New Zealand archaeological sites in the analysis (see Chapters 4 and 5). Higham (1993) used the new modelled marine calibration curve and ΔR developed by Stuiver and Braziunas (1993), and saw that the differences in calibrated age between marine shell and charcoal dates was small, the only anomalies between dates being from one site which was later found to have suffered from post-depositional disturbance (Higham pers. comm.). Higham (1993) concluded from his study that he had no preference for one sample type over another, believing that this was a site specific issue. The important aspect regarding Higham's study, was that all the charcoal used was identified to short-lived species, and the number of charcoal and marine shell radiocarbon ages used for comparison was quite large (24 charcoal and 41 marine shell). No other previous studies had such extensive and well provenanced data.

## DISCUSSION AND CONCLUSIONS

### Finding a Reliable Sample Type for Radiocarbon Dating New Zealand Archaeology

In this review, I have examined past literature concerning differences in radiocarbon dates between sample types from archaeological contexts, and the reliability of various materials for dating New Zealand prehistory. Two areas of concern are apparent. The first relates to the reliability of particular sample types for dating. Various problems identified include inbuilt age in wood, the use of bone carbonate and collagen, and potential recrystallisation of marine shell. Secondly, there is a concern about mis-association of sample and event.

It is clear from this literature review that archaeologists are divided on the reliability and usefulness of various materials available in archaeological deposits for dating (Tables 3.3 and 3.4). No particular sample type has been universally accepted either for its post-depositional stability or degree of association with the event to be dated. Can any firm conclusions be drawn from the literature as to the best material for archaeological dating? Too few studies have been made to satisfactorily answer this question, and often radiocarbon ages on a particular sample type have been rejected by archaeologists based either on studies which have been in error, or on unsubstantiated theory.

The most significant investigations that have been undertaken, were those by Trotter (1967c, 1968), Rafter *et al.* (1972), McCulloch and Trotter (1975), Law (1984), Caughley (1988), and Anderson (1989) (excluding Higham 1993, see below) where age differences between various sample types from the same archaeological contexts were compared. These comparisons were undertaken in two ways. The first method involved calculating differences in radiocarbon age between samples types from each of the archaeological sites being analysed. All the age differences from the sites were then averaged to produce an overall age difference between each sample type (Trotter 1967c, 1968; Rafter *et al.* 1972; McCulloch and Trotter 1975; Law 1984). This method took into account site history and cultural interpretation. The second method used lists of radiocarbon dates from sites which had multiple materials dated, derived a mean age from the total number of dates for each sample type, and then compared these means to find the variation in radiocarbon age for each sample type (Caughley 1988; Anderson 1989).

Both these methods of comparison produced age differences between charcoal and reservoir corrected marine shell from as great as 300 years to as little as 77 years, and between reservoir corrected marine shell and moa bone collagen, from 26 years down to 0 years. It can be seen, therefore, that both the methodologies produced very diverse results. This was because each method contained fundamental errors, which resulted in inaccurate appraisals of radiocarbon age differences between charcoal and marine shell and bone collagen sample types.

| | Charcoal | Marine shell | Bone collagen |
|---|---|---|---|
| Trotter (1967c, 1968) | N | Y | Y |
| Law (1971) | Y+ | Y | Y |
| Rafter *et al.* (1972) | N | Y | Y* |
| Grant-Taylor (1974) | Y+ | Y | Y |
| McCulloch and Trotter (1975) | N | Y | Y |
| Law (1981) | Y+ | Y | Y* |
| Anderson (1982) | Y | ? | ? |
| Law (1984) | N | Y | N |
| Trotter and McCulloch (1984) | N | Y | Y |
| Anderson (1984) | Y+ | ? | ? |
| Caughley (1988) | N | Y | Y |
| Anderson (1989) | Y+ | ? | ? |
| Anderson and McGovern-Wilson (1990) | Y+ | ? | ? |
| Anderson (1991) | Y+ | ? | ? |
| Fankhauser (1992) | N | Y | N |
| Higham (1993) | Y+ | Y | ? |

Table 3.3. Summary of opinions on the reliability of the three main samples types used for radiocarbon dating New Zealand archaeological sites. (y = reliable; y+ = reliable if twigs are used; y* = reliable but can be affected by contamination; n = not reliable; ? = reliability undecided either due to unresolved issues such as contamination, or further research required).

| | Cultural event being dated | Charcoal | Marine shell | Bone collagen |
|---|---|---|---|---|
| Grant-Taylor (1974) | Moa-hunting | N | N | Y |
| McCulloch and Trotter (1975) | Moa-hunting | N | Y | Y |
| Trotter and McCulloch (1984) | Moa-hunting | N | N | Y |
| Anderson (1984) | Moa-hunting | Y+ | N | N |
| Caughley (1989) | Colonisation of New Zealand | N | Y | Y |
| Anderson (1991) | Colonisation of New Zealand | Y+ | ? | ? |

Table 3.4. Summary of opinions on sample type reliability based on what material most likely dates the cultural event being investigated. (y = reliable; y+ = reliable when twigs are used; n = not reliable; ? = reliability undecided).

The most notable error in the investigations of Trotter (1967c, 1968), Rafter *et al.* (1972), McCulloch and Trotter (1975) and Law (1984), was that charcoal used from the sites in the comparisons was not identified to tree species. Often the observation was made that the charcoal used for dating must have been from relic wood, such as drift wood, which accounted for the charcoal dates being hundreds of years older than the marine shell or moa bone collagen ages. But because the charcoal used in these studies was unidentified, the samples dated were at high risk of being affected by inbuilt age. A second problem with these studies was the possibility that comparative samples may not have been collected from the same stratigraphic layers within a site. The only study of this kind which has been undertaken correctly where charcoal was identified to short-lived species, and the provenance of the different sample types being compared was the same, was by Higham (1993). As explained above, Higham found little difference between his paired charcoal or marine shell radiocarbon ages.

The later studies by Caughley (1988) and Anderson (1989) involved straight statistical comparisons of lists of dates. An obvious problem with these analyses was that they were general comparisons of dates without regard to the stratigraphic relationship of the dates or the site history.

Other significant factors also not considered by any of the comparative studies above (excluding Higham 1993), was the New Zealand Marine Shell Standard used by the IGNS laboratory had changed over the years, and their method to calculate reservoir corrected marine shell radiocarbon dates had also changed. This is best illustrated at Lockerbie's Pounawea site above. The only study that referred briefly to a changing marine shell standard was Anderson (1991:782). The differences in radiocarbon age between marine shell and

| Lab No. | Sample Type | Provenance | Associated Artefacts and Fauna | CRA |
|---|---|---|---|---|
| NZ-682 | Charcoal sp.? | Layer 2 | Moa bone | 728 ± 42 |
| NZ-684 | Charcoal sp.? | Layer 2 | Moa bone | 517 ± 60 |
| NZ-685 | Charcoal sp.? | Layer 2 | Moa bone | 177 ± 113 |
| NZ-683 | *Austrovenus stutchburyi* | Layer 2 | Moa bone | 1053 ± 58 |
| NZ-1250 | *Paphies subtriangulatum* | Layer 2, Shell Lens 3 | Moa bone | 695 ± 54 |
| NZ-1251 | *Austrovenus stutchburyi* | Layer 2, Shell Lens 4 | European cultural material | 1169 ± 70 |
| NZ-1347 | *Austrovenus stutchburyi* | Layer 2, Shell Lens 4 | European cultural material | 1159 ± 48 |
| NZ-1349 | *Austrovenus stutchburyi* | Layer 1, Shell Lens 1 | Moa bone | 1093 ± 52 |
| NZ-1479 | *Austrovenus stutchburyi* | Layer 2, Shell Lens 3 | Moa bone | 984 ± 56 |
| NZ-1480 | *Paphies subtriangulatum* | Layer 1, Shell Lens 1 | Moa bone | 917 ± 55 |

Table 3.5. Radiocarbon ages and provenance of samples from the Foxton moa-hunting site. (Provenance data acquired from IGNS site record forms).

other materials in early studies, therefore, may have been in error because marine shell dates compared with charcoal ages had been corrected using differing marine shell standards, or calculation methods.

This survey of the literature has identified, however, three universally accepted recommendations with respect to the dating of three sample types:

- if dating charcoal, use twigs of short-lived tree and shrub species;
- if using bone, date the collagen fraction;
- when dating marine shell, test for recrystallisation and do not use *Amphibola crenata* (mudsnail).

Of these three sample types, charcoal has received the most attention and research. Only two detailed studies, those of Anderson (1991) and Higham (1993), have been made using identified charcoal. These studies illustrated that charcoal was a reliable sample type for dating when twigs of short-lived species were used. Also, they showed that contamination of charcoal from archaeological contexts was negligible, and that the only real problem associated with this sample type was occasional inbuilt age (Anderson 1991:784-788; Higham 1993:156). Such specific individual analysis of a sample type has not been carried out on bone [11](even though there are many moa bone dates from moa-hunting sites), and the only study which looked at marine shell, paired with identified charcoal (Higham 1993), considered only three archaeological deposits, one of which had suffered from post-depositional disturbance.

Although marine shell has frequently been used to date archaeological sites in New Zealand, it has received little detailed analysis in this country (other than Higham's (1993) research). This material is considered reliable for dating at the IGNS and Waikato laboratories, as it is in overseas laboratories, but as research has been limited here in New Zealand, its short-comings, compatibility, or advantages over charcoal require further testing.

This need for further research is especially relevant when we consider common beliefs amongst the New Zealand archaeological community, about the reliability of dating particular marine shell species, and environments which may be at risk for dating.

As discussed above, Anderson (1991) outlined two main concerns regarding the radiocarbon dating of marine shell in New Zealand archaeology. First, that the estuarine bivalve *Austrovenus stutchburyi* is unreliable because it may incorporate dissolved ancient carbon within its particular habitat. This argument has been supported by one example, the Foxton marine shell radiocarbon dates. Second, that upwelling at Kaikoura, can cause marine shell radiocarbon dates from this location to appear too old. Again, the Avoca Point marine shell radiocarbon ages are the only examples to have been presented as evidence for this effect.

When we examine the radiocarbon dates and associated archaeological material from the Foxton site, post-depositional disturbance rather than the hard water effect (see Chapter 4), appears to be a more likely reason for the anomalies in the shell ages from this site (Table 3.5; see Figure 3.6).

[11] Dr Fiona Petchey (1998) has recently completed research on the reliability of Snapper (*Pagrus auratus*) and Barracouta (*Thyrsites atun*) bone for radiocarbon dating New Zealand archaeological sites, and has identified these fish species as reliable for dating.

| Lab. No. | Marine Shell Species | Provenance | CRA |
|---|---|---|---|
| NZ-2718 | *Lunella sp.* | Square M20, 10-20 cm depth, occupational deposit. | 1183 ± 29 |
| NZ-2719 | *Protothaca sp.* | Square M20, 10-20 cm depth, occupational deposit. | 1174 ± 33 |
| NZ-6472 | *Haliotis iris* | 40m, Trench A, 55-105 cm depth. Beach gravels with shells and bones and stone flakes (cultural) in black silt. | 871 ± 40 |
| NZ-6525 | *Cookia sulcata* | Occupational deposit outside wall area relating to early occupation of area. | 800 ± 32 |

Table 3.6. Radiocarbon ages from archaeological sites excavated at 'Avoca Point'.

If we assume the young charcoal age NZ-685 has not suffered from inbuilt age (none of the charcoal samples were identified to species prior to dating), it appears anomalous given its association with the moa bone. When the marine shell radiocarbon ages are examined, the dates of *Austrovenus stutchburyi* for both layers of the site are not unusual, compared to shell $^{14}C$ determinations from other moa-hunting sites such as Shag Mouth or Pleasant River (see Higham 1993). The *Paphies subtriangulatum* radiocarbon date NZ-1250, however, appears to be too young given its association with moa remains (as found with its paired modern charcoal age). The *Austrovenus stutchburyi* radiocarbon ages NZ-1251 and NZ-1347 are typical for a moa-hunting site, but these are found alongside European material.

The radiocarbon anomalies identified at the Foxton site occur at the interface of lenses 3 and 4 in Layer 2 (see Figure 3.6). These lenses have characteristics which indicate mixing of the two deposits; the young *Paphies subtriangulatum* and charcoal radiocarbon ages from Lens 3 of Layer 2 found with moa remains, and the older *Austrovenus stutchburyi* radiocarbon dates alongside European cultural material in Lens 4 of the same layer. In short, there are severe qualifications which have to be made regarding the interpretation that an environmental factor has caused the variable radiocarbon ages from the Foxton site. It is more likely that the errors are related to post-depositional mixing of cultural deposits. Further research would be required to confirm this.

The marine shell radiocarbon dates from Avoca Point are more easily explained than the Foxton radiocarbon ages. NZ-2718 and NZ-2719 from the site are *ca.* 300 years older than NZ-6472 and NZ-6525 (Table 3.6). However, the sample record forms from the IGNS laboratory indicate that not only were samples NZ-2178 and NZ-2179 collected ten years earlier than NZ-6472 and NZ-6525, but both sets of radiocarbon dates came from differing locations. In addition, the provenance of the samples dated from each occupational deposit are at markedly differing depths, and both collection sites are known by differing names, NZ-2718 and NZ-2719 come from Old Pier Point, and NZ-6472 and NZ-6525 from the Fyffe's site.

The radiocarbon dates, therefore, appear to date two different occupational deposits of varying age. The old dates are not caused by upwelling as Anderson (1991:777) and Caughley (1988) have suggested. Higham (1993:137-138) undertook a small study regarding upwelling at Kaikoura by dating modern *Austrovenus stutchburyi* taken close to the Fyffe's moa-hunting site, and found that the sample had not been affected by upwelling. Because only one sample was tested, Higham concluded that further research concerning upwelling at the site would be useful (1993:138).

The Foxton and Avoca Point examples illustrate, therefore, that the anomalous radiocarbon ages on the shellfish species dated may not be due to the reasons originally proposed. In addition, they provide further evidence of the need for testing the reliability of marine shell to establish its usefulness for radiocarbon dating New Zealand prehistory.

## A Solution to the Sample Type Dilemma

It is apparent that an investigation into the reliability of marine shell for New Zealand archaeology would be very useful at this time. As charcoal and marine shell are ubiquitous in New Zealand prehistoric sites, investigating the use of marine shell would greatly increase confidence in the choice of samples to date from archaeological sites.

The most satisfactory method for evaluating the reliability of marine shell is to use marine shell/identified charcoal pairs. An alternative to charcoal would be moa egg shell from an archaeological context, as Higham's (1993; 1994) research has found this material to be reliable for dating. However, moa egg shell is not widespread in northern New Zealand archaeological sites (Higham 1993:96; 1994:168).

This study will investigate the use of marine shell in radiocarbon dating New Zealand prehistory, by comparing charcoal radiocarbon ages with marine shell dates from archaeological sites sampled from various locations around New Zealand.

However, before a comparison can be made between these sample types, it is vital to recognise factors which can influence the accuracy of radiocarbon dates from these materials. For example, how does hard water affect the radiocarbon dating of marine shells, and what factors still affect charcoal dates even when the charcoal has been identified to twigs of short-lived species prior to dating? In addition, the way in which marine shell CRAs are calibrated must be scrutinised to determine whether this provides any significant source of error.

# CHAPTER 4

# RADIOCARBON DATING CHARCOAL AND MARINE SHELL: CONSIDERATIONS FOR USE

## INTRODUCTION

To test properly the reliability of a sample type for radiocarbon dating an archaeological site, particular considerations must be taken into account. This chapter explains what factors can affect the true radiocarbon ages of charcoal and marine shell samples used to date New Zealand archaeological sites.

## THE PROBLEM OF INBUILT AGE

Because radiocarbon dating aims to measure the time lapsed since an organism's death, inbuilt age of charcoal can make dating of archaeological contexts problematic. In an attempt to alleviate this problem, archaeologists have, since 1976, identified the charcoal recovered from archaeological contexts to short-lived tree and shrub species. Selecting twigs additionally refined this sample where it was assumed that these represent only a few years of growth, and such delicate foliage would not survive long after death (see Chapter 3).

Anderson (1991) and McFadgen *et al.* (1994) have tackled the problem of inbuilt age in using charcoal to date New Zealand prehistory. Anderson (1991) found that even though charcoal may be identified to twigs of what is believed to be short-lived species, some of these species may in fact live considerably longer than assumed. For example, he noted that some botanists believe the *Pseudopanax* species and kanuka *(Kunzea ericoides)* may live up to 200-300 years, and that manuka *(Leptospermum scoparium)* may live up to 100 years. Anderson also found that New Zealand shrubs can probably live for 50-100 years (Anderson 1991:780). He saw that the problem of inbuilt age was enhanced for samples being used to date sites of brief occupation, and that "Charcoal samples of 'short lifespan' species, even where the material also appears to be of young wood, still indicate inbuilt ages of 50-150 years, or greater..." (Anderson 1991:781).

In dating the colonisation of New Zealand using charcoal, McFadgen *et al.* (1994:224) separated samples radiocarbon dated using short-lived species (< 100 years) from those on long-lived species (>100 years), based on the criteria shown in table 4.1. They observed that when separating charcoal ages using this criteria, there were fewer dates on short-lived species with ages > 700 years cal BP when compared to dates on long-lived species (McFadgen *et al.* 1994:224-225).

| Short (< 100 yr) | Medium (100–300 yr) | Long (> 300 yr) |
|---|---|---|
| *Aristotelia serrata* | *Ackama rosifolia*† | *Agathis australis* |
| *Brachyglottis* sp. | *Alectryon excelsus* | *Dacrydium cupressinum* |
| *Carmichaelia* sp. | *Beilschmiedia* sp.‡ | *Halocarpus kirkii* |
| *Carpodetus serratus*‡ | *Cordyline australis* | *Lagarostrobus colensoi*† |
| *Cassinia* sp. | *Corynocarpus laevigatus* | *Laurelia novaezelandia*† |
| *Coprosma* sp. | *Discaria toumatou* | *Libocedrus bidwillii*† |
| *Coriaria* sp. | *Dysoxylum spectibile* | *Metrosideros* sp. |
| *Corokia macrocarpa* | *Hoheria* sp.† | *Nothofagus* sp.† |
| *Geniostoma rupestre* | *Knightia excelsa* | *Phyllocladus* sp.† |
| *Hebe* sp. | *Kunzea ericoides* | *Podocarpus totara* |
| *Hedycarya arborea*‡ | *Myrsine divaricata*† | *Prumnopitys spicatus* |
| *Leptospermum scoparium*‡ | *Myoporum laetum* | *Vitex lucens*† |
| *Leucopogon fusciculatus* | *Nestigis* sp.‡ | |
| *Lophomyrtus obcordata*‡ | *Olearia* sp. | |
| *Macropiper excelsus* | *Pseudopanax* sp.† | |
| *Melicytus ramiflorus*‡ | *Paratropis microphylla* | |
| *Melicytus* sp.‡ | *Pittosporum eugenoides* | |
| *Myrsine australis*‡ | *Pittosporum tenuifolium*† | |
| *Myrsine* sp.‡ | *Plagianthus* sp. | |
| *Olearia rani*‡ | *Sophora microphylla* | |
| *Pseudopanax arboreus*‡ | *Sophora* sp. | |
| *Pseudopanax crassifolius*‡ | *Weinmannia* sp. | |
| *Pseudowintera* sp. | | |
| *Pteridium esculentum* | | |
| *Schefflera digitata* | | |
| Tree fern | | |

*Data provided by Dr. Philip Simpson, Botanist, Science and Research Division, Department of Conservation.
†Life span can be much shorter than designated years.
‡Life span can be longer than designated years.

Table 4.1. McFadgen *et al's.* (1994:224) table of life expectancy of plant species used for radiocarbon dating in New Zealand.

| Tree and Shrub Species | | Life Expectancy (McFadgen *et al.* 1994:224; Simpson pers. comm.) |
|---|---|---|
| **Latin Name** | **Common Name** | |
| *Agathis australis* | Kauri | l |
| *Alectryon excelsus* | Titoki | m |
| *Avicennia marina var. resinifera* | Manawa/Mangrove | s* |
| *Beilschmiedia tarairi* | Tarairi | m* |
| *Carpodetus serratus* | Putaputaweta | s* |
| *Coprosma sp.* | | s |
| *Dacrydium cupressinum* | Rimu | l |
| *Dodonaea viscosa* | Akeake | *m |
| *Dysoxylum spectabile* | Kohekohe | m |
| *Hebe sp.* | Koromiko-taranga | s |
| *Hedycarya arborea* | Porokaiwhiria/ Pidgeonwood | s* |
| *Kunzea ericoides* | Kanuka | m |
| *Leptospermum scoparium* | Manuka | s* |
| *Litsea calicaris* | Mangeao | m |
| *Lophomyrtus bullata* | Ramarama | s |
| *Lophomyrtus obcordata* | | s* |
| *Macropiper excelsum* | Kawakawa/Pepper tree | s |
| *Melicytus ramiflorus* | Mahoe | s* |
| *Metrosideros excelsa* | Pohutukawa | l |
| *Metrosideros robusta* | Vine Rata | l |
| *Myrsine australis* | Mapau | s* |
| *Olearia sp.* | | m |
| *Pittosporum sp.* | | *m |
| *Plagianthus regius* | Manatu/Ribbonwood | m |
| *Podocarpus totara* | Totara | l |
| *Prumnopitys taxifolia* | Matai | l |
| *Pseudopanax arboreus* | Puahou/Fivefinger | s* |
| *Pseudopanax colensoi* | Orihou | *m |
| *Pseudopanax crassifolius* | Horoeka/Lancewood | s* |
| *Pseudopanax sp.* | | *m |
| *Pseudowintera colorata* | Horopito | *m |
| *Pseudowintera sp.* | | s |
| *Pteridium esculentum* | Bracken fern | s |
| *Schefflera digitata* | Patete | s |
| *Vitex lucens* | Puriri | *l |
| *Weinmannia racemosa* | Kamahi | m |

Table 4.2. Tree and shrub species commonly used for radiocarbon dating New Zealand prehistory as seen in Waikato and IGNS laboratories sample record forms. Life expectancy are based on McFadgen *et al.* (1994:224) data and information provided by Dr. P. Simpson. s= short (<100 years); m=medium (100-300 years); l=long >300 years). * to left of life expectancy indicates plant frequently lives shorter than this time, * to the right means the plant occasionally lives longer than this time.

Table 4.2 presents tree and shrub species which have been commonly used for radiocarbon dating in New Zealand archaeology since 1976, as seen in sample record forms from Waikato and IGNS radiocarbon dating laboratories. The life expectancy of these trees based on McFadgen *et al.'s* (1994: 224) estimates can be seen in column 3. The majority of species are in the short to medium age range, with a few coming from long-lived species. However, these floral lifespans do not take into account that the samples dated were from twigs, and sometimes from seeds, so although tables 4.1 and 4.2 give an indication of the maximum life span of the flora being dated, the actual age of the wood is related to the number of years of growth the sample represents. So what do Anderson (1991) and McFadgen *et al.'s* (1994) studies (and Table 4.2), tell us about the inbuilt age of charcoal used to date New Zealand archaeological sites? Though both agree that twigs of short-lived species should be used, two areas of debate arise: 1) some traditionally accepted short-lived tree and shrub species may in fact have been misidentified, and live for years well in excess of what is believed, and 2) if twigs only represent a few years of growth, then twigs from long-lived species may still be acceptable for dating. McFadgen *et al.'s* (1994) short-lived and long-lived species radiocarbon age comparison does not indicate whether samples of twigs only were analysed. If we consider Anderson's (1994) observations that some short-lived species (such as the *Pseudopanax* sp.) may live for 200-300 years, then some of the radiocarbon samples in McFadgen *et al.'s* (1994) 'short-lived' category may in fact have high inbuilt age, because either the material was not derived from twigs, or the tree species may not actually be short-lived. Alternatively, some of McFadgen *et al.'s* (1994) long-lived material may come from twig material which represent only a few seasons growth.

The problem of inbuilt age, then, may not be fully resolved by selecting twigs from short-lived species because, as Anderson (1991:780) noted, it is difficult to obtain data on longevity of New Zealand trees and shrubs. It is, however, the most effective solution at present to resolve the inbuilt age problem.

One aspect of inbuilt age not often closely examined, and which may have a greater influence on dates on charcoal samples than believed, is the storage age or durability of tree and shrub foliage (see Anderson 1991:780-781). When selecting suitable charcoal for dating, the selection of twigs of short-lived species is believed to account for both the growth and storage age (as discussed above). It is often presumed that by selecting a twig, storage age is immediately accounted for, and the only remaining area of concern is the growth age. This is true when attempting to exclude ancient driftwood from a sample where twigs are less likely to survive in a coastal environment because of their fragility. However, trees which have died and lie in forests or on forest margins, may vary in both their durability and time taken to decay in the environment. It is generally believed that most short and medium-lived trees and shrubs decay within 100 years in the New Zealand environment (Simpson pers. com.[12]). The long-lived species such as totara, puriri, and kauri, decay very slowly, but beech, also a long-lived species, decays quickly. Decay rate is also therefore species specific. Very little information is available on the decay rate of New Zealand trees in either forest or coastal environments.

Clifton (1990:4-5) undertook a test commonly used by commercial timber industries to provide "natural durability ratings" of New Zealand indigenous timbers. The test involves placing 50 x 50 mm stakes from the heartwood of a variety of tree species in the ground, and then once a year the stakes are knocked on two adjacent sides using a wooden mallet (Clifton 1990:4). If the stakes do not fail then they are dug up and a rating of decay is given. However, Clifton (1990:5) observed that these indicators are "comparative and not absolute" because of the wide range of results. Also, that the ratings come from tests carried out in seven different regions around New Zealand and are therefore influenced by a variety of climates, topographies and soil types (Clifton 1990:5).

Factors such as this can dramatically affect the durability and decay rates of timbers. For example, Simpson (pers. comm.) and Anderson (1991:780) both note that in dry environments such as in Central Otago, trees may take hundreds of years to decay. Alternatively, Barber (pers. comm.[13]) observes that because of the infertile nature of the Nelson/Marlborough area, beech will not grow as large as in other locations of the South Island, nor for long periods of time, and the environment brings about rapid decay after death. Forest wood used by people in prehistory most probably came from the surface of the ground where the tree fell. Clifton (1990:4) noted that "a species ranked moderately durable in the ground would be classified as durable above the ground". Hence durability and decay tests on New Zealand timbers at present will only give a general indication of the time a fallen tree will lie on the forest floor relatively intact and be available as firewood. Because of this, the influence of decay rates on inbuilt age of charcoal samples can not currently be determined.

## LIFEWAYS AND HABITATS OF SHELLFISH

The two classes of mollusc commonly used to radiocarbon date archaeological sites are bivalves and gastropods. To understand the pattern of $^{14}C$ uptake in shellfish, it is necessary to examine the way in which the shell is constructed and the habitats in which the differing species dwell. From this information the sources of $^{14}C$ in the calcium carbonate of the shell may be identified, and the influence of shellfish habitat and environment on the radiocarbon date examined.

### Construction of the Shell and Carbon Sources

The shell of bivalves and gastropods have essentially the same structure and are formed in a similar manner. The shell comprises a periostracum cover under which lies between 2 to 4 calcareous layers (Figure 4.1). The main function of the periostracum is to aid shell secretion and, additionally, to protect the calcium carbonate below from dissolution (Barnes 1987:403). The calcareous structure may be totally aragonitic, (eg. cockles, pipi, tuatua or whelks), or calcitic (eg. oysters). The calcareous layers are deposited either as prisms, minute laths, or tablet structures in the form of sheets, lenses or more complex forms (Barnes 1987:404). The basic building blocks of the shell are the prism and tablet structures which are laid down within an organic matrix. These structures along with the periostracum, can account for up to 72% of the shell's dry weight. The mantle, which overhangs the body of the mollusc beneath the shell, forms a large sheet of tissue located under the valves and consists of three folds; an inner, middle and outer (Figure

[12] Dr. P. Simpson, Botanist, Science and Research Division, Department of Conservation, Wellington.

[13] Dr. I. Barber, Archaeologist, New Zealand Historic Places Trust, Wellington.

4.1). The inner fold contains the radial and circular muscles of the shellfish and are used for movement. The middle fold has a sensory function. The outer fold is active in shell secretion with its inner surface laying down the periostracum, and then the outer surface secreting the first calcareous layer (Barnes 1987:404). The whole mantle surface then secretes the rest of the calcareous portion of the shell. Shell secretion begins in the small gap between the mantle and the shell. This gap contains the extrapillial fluid from which is derived both the calcareous elements and the surrounding organic framework consisting of the protein conchiolin (Figure 4.2 and 4.3) (Barnes 1987:350, 404-405).

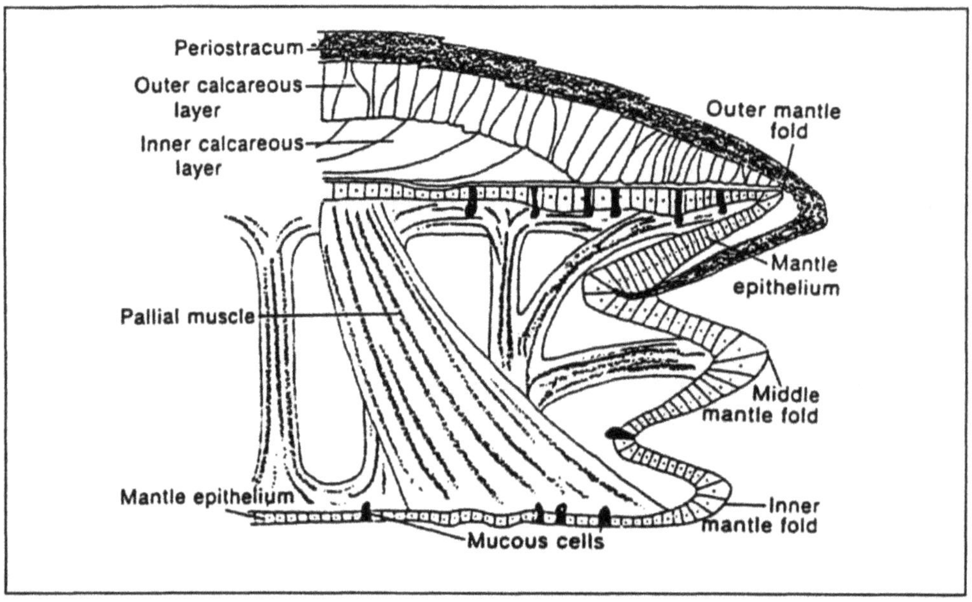


Figure 4.1. Transverse section through a bivalve showing calcareous layers, mantle folds and the periostracum, which aids in shell secretion (from Barnes 1987:403).

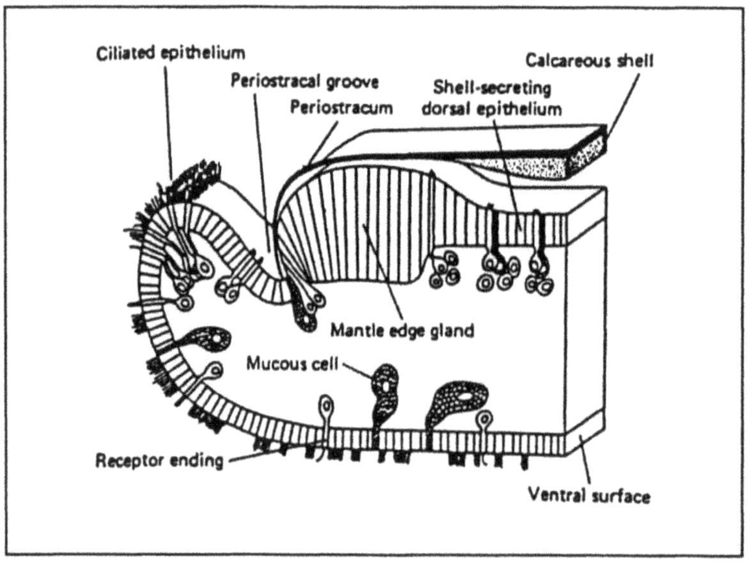


Figure 4.2. Shell secretion in molluscs as seen in a pulmonate snail. Barnes (1987:351) explains: "Periostracum is secreted initially at the bottom of the periostracal groove and is thickened as it passes over the mantle edge. Secretion of the calcareous portion of the shell begins to the inner side of the mantle edge gland.".

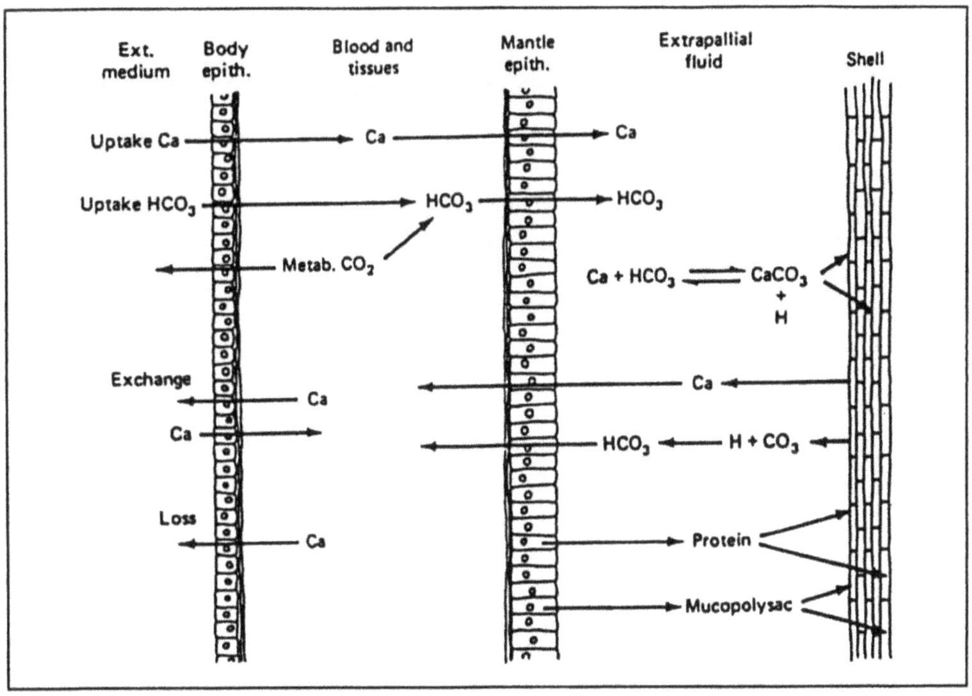


Figure 4.3. Process by which shell is formed by bivalves and gastropods. Barnes (1987:405) explains: "Protein and mucopolysaccharides are involved in the production of the organic framework of the shell. Calcium loss occurs during periods of dissolution".

The elementary carbon which forms the $CaCO_3$ of the shell comes from three sources: DIC (dissolved inorganic carbon) in the form of $HCO_3^-$ from the seawater, contributing the largest proportion of carbon; metabolic carbon in the form of $CO_2$ from terrestrial and marine organic food sources; and possibly a minor amount from atmospheric $CO_2$ for estuarine species (Tanaka *et al.* 1986:520; Little 1993:468; Higham 1993:121-123). Bivalves, such as *Austrovenus stutchburyi* (cockle), *Paphies australis* (pipi) and *Paphies subtriangulatum* (tuatua), draw seawater and suspended organic matter through an inhaling siphon over paired ctenidia (gills). The ctenidia act as gas exchange media as well as filters separating debris from food (Barnes *et al.* 1993:133-134). From here carbon is taken both directly into the mantle from the surrounding water (as bicarbonate ions), as well as from the digestion of organic matter (metabolic carbon in the form of $CO_2$) (Figures 4.2 and 4.3). The Gastropods *Lunella smaragda* (catseye), *Cominella adspersa*, and *Cominella virgata* (the later whelks), also obtain their carbon from the immediate seawater, but their metabolic carbon comes from the use of radula in *Lunella smaragda*, or extensible proboscis feeding for *Cominella* sp., rather than from filtering. *Lunella smaragda* graze upon organic deposits and corallines (algae), and both *Cominella* sp. are carnivorous, (though *Cominella adspersa* is a scavenger rather than a predator) (Morton and Miller 1968:391, 398).

Marine shell radiocarbon ages are related to the intake of DIC from the marine environment, but this may be affected by the incorporation of metabolic carbon. This and other factors are dependent on the habitat and environment of the mollusc species being dated, and therefore a knowledge of species habitat is important (see Tanaka *et al.* 1986; Little 1993; Higham 1993; Dye 1994).

## HABITATS OF SHELLFISH USED FOR RADIOCARBON DATING NEW ZEALAND ARCHAEOLOGICAL SITES

### Bivalves

The most common shellfish found in Maori prehistoric middens are *Austrovenus stutchburyi*, *Paphies australis* and *Paphies subtriangulatum*. *Austrovenus stutchburyi* lives in a very similar habitat to *Paphies australis*. Both species are found on broad level beach flats either in sheltered bays or at the mouths of estuaries, which are protected from harsh wave environments (Morton and Miller 1968:443). Pipi prefer a coarse sand to inhabit, though will extend up onto the edges of sheltered mudflats of enclosed beaches. Morton and Miller (1968:486) note that pipi "flourish" in this habitat when fresh water seeps up through the beach. Cockles can also be found in this habitat as well as on the enclosed beach mudflats. Both the protected and enclosed beach areas experience gentle wave movement only, which aids in the stirring up of silt for filter feeding. These shellfish are also typically immobile and are found in clusters, though cockles do occasionally move along the beach to more preferable feeding areas. Pipi and cockles can be found burrowed just below the surface up to 4 cm deep.

*Paphies subtriangulatum* are typical of the open beach habitat where clean sand and direct wave action are constant in an unsheltered surround (Morton and Miller 1968:443). These shellfish are always moving with the shifting sands and can be found, as with pipi, only a couple of centimetres below the surface. *Venerupis largillierti*, the oblong cockle, is also sometimes found in middens. This species inhabits silty sand of enclosed beaches often in association with the soft shore plants of the *Zostera* sp., which lie at the foot of mangrove swamps (Morton and Miller 1968:543). This species of bivalve burrows down 5 to 6 cm into the sand from where it inhales food and water with its long siphons which extend to the surface.

Of the rocky shore bivalve species, *Mytilus edulis aoteanus* (blue mussel), *Perna canaliculus* (green mussel), and *Crassostrea glomerata* (the rock oyster) were commonly exploited shellfish in prehistory. The blue mussel was harvested in the southern areas, whereas the green mussel was taken in the northern region. Both species are common in their particular provinces, but less common in proportion to each other on their opposing islands. *Perna canaliculus* usually inhabits open coasts but is also found in sheltered coastal locations (Morton and Miller 1968:62, 304, 342-343). This mussel attaches itself to rocks using tough byssus threads, and forms masses of individuals which provide more resistance to pounding waves of the open coast. *Mytilus edulis aoteanus* inhabits a more sheltered environment, only appearing occasionally on open coasts and then only in small clusters (Morton and Miller 1968:304, 342-343). Their preferred habitat are inlets with a strong passing water current where, as with *Perna canaliculus*, they are attached to the rock strata with strong byssul threads, or in pools behind the splash zone of the open coast. When in association with *Mytilus edulis aoteanus* in sheltered environments, *Perna canaliculus* will always be seen living on the rock faces immediately below.

*Crassostrea glomerata* cements itself to hard, erosion resistant rocks, preferring vertical and inclined surfaces, lying either flat against the rock when found in distributed numbers, or forming oyster clusters in mass. They prefer a sheltered environment, such as in bays or along rocky harbours, good exposure to sunlight, and are found attached to rocks above the level of settling silt (Morton and Miller 1968:406). This species flourishes in harbours and thins in number as wave action increases (Morton and Miller 1968:64).

## Gastropods

Various gastropods were also exploited by prehistoric Maori, such as *Lunella smaragda*, *Cominella* sp., and *Amphibola crenata* (mudsnail). All these species have been used to radiocarbon date archaeological sites in New Zealand. *Amphibola crenata* is no longer used for dating due to its unreliability (see Chapter 3). *Lunella smaragda* are found most often in tidal pools and on the surface of rocky shore platforms where it can graze on algae and kelp (Morton and Miller 1968:202-203). This habitat is moderately sheltered with the rocky crevices and pools providing cover from wave action. Although *Lunella smaragda* can be present in muddy areas (such as amongst *Zostera* sp. and in mangrove swamps), they prefer rocky strata. *Cominella adspersa* live in a number of niches, from rocky shore, to muddy sand and sandy beaches, as long as the harbour, bay or estuary is sheltered (Morton and Miller 1968:398,490). Here they scavenge dead organisms. *Cominella virgata*, more predator than scavenger, inhabits a moderately sheltered niche living on rocky or silted reefs, and can be found in tidal pools under stones. As with *Cominella adspersa*, *Cominella virgata* can move quickly between the tides (Morton and Miller 1968:85, 391).

# FACTORS THAT INFLUENCE THE RADIOCARBON DATING OF SHELLFISH

## The 'Hard Water Effect' and 'Upwelling'

Each of these factors are common explanations for variations in marine shell ages. They cause marine shell radiocarbon dates to appear too old. This is caused by an increase in the amount of $^{14}$C-depleted dissolved inorganic carbon (DIC) (also known as dead or old carbon), within the hydrological environment in which the shellfish live. This $^{14}$C-depleted carbon dilutes the natural levels of $^{14}$C in the water, decreasing $^{14}$C activity, and therefore increasing the age of the shellfish.

### *The Hard Water Effect*

The 'hard water effect' can originate from two sources. Water drainage from geological landscapes of limestone or other highly calcareous strata, can introduce $^{14}$C-depleted carbon into a marine reservoir, where it is then incorporated into the shells of marine, estuarine or lacustrine molluscs. This effect is also caused when shellfish live upon calcareous strata which becomes a source of $^{14}$C-depleted carbon within the surrounding waters. The hard water effect is usually associated with estuarine and lacustrine shellfish species (see below). *Austrovenus stutchburyi* has been proposed as potentially at risk of being affected by hard water in New Zealand (McFadgen 1972:54, 1978a:112).

The hard water effect has been observed in many overseas studies on the dating of marine shell, particularly from lacustrine environments. Dye (1994:55) has found greater age discrepancies between shellfish species inhabiting a limestone coast, than those from non-calcareous geological deposits. Berger and Meek (1992) noted that radiocarbon dates of the freshwater mollusc *Anodonta californiensis* from the Humboldt River basins in Nevada, require an 1870 ± 270 year reservoir correction due to an influx of dead carbon from dissolved Paleozoic carbonates through which the river flowed. This was in contrast to the same genus being dated in the Mojave River basins, southern California, where the basin sits upon a carbonate free crystalline bedrock, therefore requiring a reservoir correction of only 450 years.

A hardwater effect is also noted by Heir-Nielsen *et al.* (1995) for Danish Fjords. In the Limfjord Sound in particular, the reservoir ages were greatly influenced by DIC influxing from ground water and streams. The fjords studied showed reservoir ages spanning 400 to 900 years, depending upon local hydrological and geological conditions (Heir-Nielsen *et al.* 1995). Aravena *et al.* (1991:170) observed the influence of "older water" on the lowering of $\Delta^{14}$C activities in a fresh water lake in the Precambrian Shield in Canada.

Bronic *et al.* (1991:181-182) compared $^{14}$C activities of pre-bomb known age carbonate deposits and wood samples to determine the magnitude of the hard water effect in lake sediments in the Plitvice Lakes area of Yugoslavia. Here the $^{14}$C activity of wood fragments in lake sediments were compared with the activity of surrounding lake sediments to determine the initial $^{14}$C activity of the lakes, and therefore generate a correction factor for future lake sediment age determinations.

Fontes *et al.* (1991:196-197) observed the influence of "old carbon" on radiocarbon ages of the freshwater mollusc *Cardium* from Holocene lacustrine deposits in a confined acquifer in the Northern Sahara. The ages were 2000 years older than expected, with the old carbon coming from "vertical seepage" from muds at the base of the acquifer. Garcia *et al.* (1991:198) account for the difference of radiocarbon ages between geological deposits of carbonates and organic matter from a paleolake as being due to a hard water effect. In avoiding the hard water effect when dating laminated lake sediments, Hajdas *et al.* (1991:205-206) used the AMS technique to date macro fossils such as leaves, needles and seeds from trees. This was undertaken in a small Swiss freshwater lake on a sequence of lake sediments ranging from 6,000-12,000 years in age (Hajdas *et al.* 1991:206).

*Upwelling*

Upwelling occurs when $^{14}C$-depleted carbon from the deep ocean is brought to the mixed layer ocean surface through upwelling sea currents, often in regions with plunging coastal marine shelves. The depleted carbon may also be upwelled from dissolving limestone sources near to rising sea currents. This factor mainly affects open marine and rocky shore shellfish species. Kaikoura in the South Island has been argued as a possible risk area for upwelling in New Zealand (see Anderson 1991:777).

The California coastline is a particularly problematic region for upwelling, where highly depleted $\Delta^{14}C$ levels have been observed by Druffel and Williams (1991), Stuiver and Braziunas (1993:153), Flessa *et al.* (1993) Ingram (1994) and Flessa (pers. comm.[14]). This coastline has seen dramatic shifts in reservoir ages, ranging from between 200 yrs to 500 yrs. This phenomenon has also been investigated by Rasmussen and Rahbek (1993) for Greenland, Gordon and Harkness (1993) for the Antarctic, Glynn *et al.* (1983) for the Gulfs of Tehuantepec, Papagaya and Panama on the Pacific coast of Central America, Kashgarian *et al.* (1994) for the Indonesian seaway, Weidman and Jones (1993) for Georges Bank, Florida, Southon *et al.* (1994) for the Indian Ocean, and by Druffel and Suess (1983) for the Peru Current. It has been suggested that upwelling may be geographically and temporally variable due to seasonal and other natural events, such as the Monsoon season (Higham 1993:137-138; Kashgarian *et al.* 1994).

## The Organic Carbon (OC) Effect: Alterations In $\delta^{13}C$ Levels

The influence of terrestrial organic matter on the age of shellfish can cause dates to be either too old or too young. Contribution by organic carbon can be detected by low $\delta^{13}C$ levels, unlike contamination by the hard water effect or upwelling. The source of organic carbon comes from terrestrial organic matter such as plant detritus, diatoms, protozoa, kelp, phytoplankton, humus or peat, carried by rivers into an estuarine environment (Little 1993:468; Higham 1993:121). This organic carbon, also known as metabolic carbon, is incorporated into the shell through nutrition. Depending upon the origin of the terrestrial material, shell ages may be increased or decreased.

An increase in shell age comes about when the organic matter fed upon in the estuary originates from an ancient source depleted in $^{14}C$, such as peat. In this case, the $\delta^{13}C$ level tends to become more negative. Conversely, organic matter which is contemporary with the shellfish during its growth may result in shell ages which are too young, with the $\delta^{13}C$ level also more negative (Little 1993:468; Higham 1993:145-146). This is because the organic matter is of terrestrial origin, and so not effected by the oceanic reservoir effect. Hence the age of the contaminating organic carbon will be *ca.* 400 years younger than the contemporary marine shell. The OC effect is possibly species specific according to Higham (1993:144-150;152-153), as mollusc habitat will determine degree of potential exposure to this influence.

## DISCUSSION AND CONCLUSIONS

Charcoal used to radiocarbon date New Zealand archaeological sites has been identified to twigs of short-lived species since 1976 to reduce the influence of inbuilt age. However, a lack of data on the actual longevity of some tree and shrub species, and deciding whether a twig from a long-lived tree species should be acceptable as datable charcoal, still raises concerns when dating this sample type. An additional concern is the durability and decay of various tree and shrub species after death. As durability and decay are species- and environment-dependant, certain regions of New Zealand may have had more sources of old wood available for prehistoric activities than others. The risk of dating twig wood of high inbuilt age is possibly greater, therefore, for archaeological sites in central Otago in the South Island, than it is for sites in the Nelson/Marlborough district. As this area of inquiry has not been studied in detail in New Zealand, the effect of this source of error on charcoal radiocarbon ages is difficult to quantify at present.

Most of the chronological problems observed with the use of charcoal come from sites which have only a single charcoal age (as seen in Chapter 3). Again, not all of the charcoal was identified in these studies. Higham (1993) and Anderson *et al.'s* (1996) extensive dating of the Shag Mouth moa-hunting site, are the best examples in New Zealand archaeology at present of the effective use of charcoal in radiocarbon dating. In these studies, multiple samples were identified to twigs of short-lived species prior to dating, the resulting dates closely agreeing with associated marine shell ages. By undertaking multiple dating, charcoal samples affected by inbuilt age would stand out from the remaining pool of dates, thus the charcoal chronology itself was a check on inbuilt age.

By investigating the origins of carbon used by bivalves and gastropods during shell building, and understanding their habitats and lifeways, we may identify what environmental and geological factors may influence the radiocarbon dating of shellfish commonly found in New Zealand archaeological sites.

For the soft shore harbour and estuarine bivalve species *Paphies australis, Austrovenus stutchburyi* and *Venerupis largillierti,* living in both protected and enclosed beaches, the two main risks come from the hard water effect and

[14] Dr. K. Flessa, Department of Geosciences, University of Arizona, Tucson, USA.

terrestrial organic carbon sources. The soft shore beach open coast species, such as *Paphies subtriangulatum*, are probably only at risk of being effected by upwelling due to their isolated habitat in clean sands and exposure to unsheltered wave action.

The rocky shore harbour and open coast species *Mytilus edulis aoteanus*, *Perna canalicula*, *Crassostrea glomerata*, *Lunella smaragda*, *Cominella adspersa*, and *Cominella virgata* may all be affected by hardwater originating directly from dissolution of the rock strata upon which they live. High risk areas would be limestone rocky shores and highly calcareous sediments. Additionally, upwelling may also be a source of old carbon where inhabited rocky shelves have a plunging coastline or offshore ocean trench.

Potential sources of error in dating charcoal and marine shell from archaeological sites have been presented above. If an archaeological site sampled for dating has not been post-depositionally disturbed, then these factors may be proposed as possible reasons for disagreements between charcoal and marine shell radiocarbon ages. The next chapter examines another consideration in the use of marine shell for dating, the marine reservoir correction.

# CHAPTER 5

# RESERVOIR CORRECTIONS

## INTRODUCTION

The calibration of marine shell conventional radiocarbon dates has changed dramatically over the past 14 years. Prior to 1986, marine shell CRAs were calibrated by subtracting a country-specific reservoir correction *(R)*, and calibrating the resulting age using an atmospheric calibration curve. Stuiver *et al.* (1986) then developed an alternative method of calibration, through the use of a modelled marine curve and application of a local reservoir correction called ΔR. This chapter describes how marine shell CRAs are calibrated using the reservoir correction *(R)* technique, and the more recent Stuiver *et al.* (1986, 1993) modelled marine curve calibration method. Problems associated with the application of each method are presented, as well as a detailed review of various studies on ΔR since its introduction in 1986.

## THE RESERVOIR EFFECT AND RESERVOIR CORRECTIONS

After $^{14}$C is produced in the atmosphere, its mixing and concentration in the various reservoirs is not equal. Replenishment of $^{14}$C in the mixed layer ocean (<75m depth) and deep ocean (>75m) is slow compared to that of the atmosphere. Whereas the "residence" time of a carbon atom is about 5 years in the atmosphere, the deep ocean can contain carbon atoms which have been present for thousands of years. The mixed layer ocean contains a $^{14}$C concentration that is intermediate between higher atmospheric levels and depleted deep ocean water. This results in the mixed layer ocean having an apparent $^{14}$C age of *ca.* 400 years, and is known as the reservoir effect (Aitken 1990:64). The reservoir effect therefore causes marine shell ages to appear *ca.* 400 years older than contemporary terrestrial material.

Radiocarbon laboratories worldwide have calculated local reservoir corrections which are applicable to marine shell radiocarbon ages for their particular region of ocean (see Stuiver and Polach 1977:356-357; Deevey 1984:3,5). These location specific corrections are then applied to their marine shell radiocarbon dates before calibration. The reservoir correction R(t)[15] is derived in three ways, either by:

- radiocarbon dating a marine shell sample of known year of collection (preferably shell collected before 1950) after correction for the industrial effect;
- comparing a marine shell radiocarbon age collected in calendar year 't' with an atmospheric radiocarbon age taken from a tree ring record of the same year of collection;
- or by comparing the age difference between paired charcoal and marine shell radiocarbon dates from the same archaeological provenance.

To calculate R(t), each method uses the formula:

$$R(t) = P - Q$$

where P is the marine shell radiocarbon age, and Q is either the known calendar age of collection, or the appropriately corrected comparable terrestrial radiocarbon age.

The reservoir correction R(t) is specific to the calendar year for which it has been determined. To simplify the correction of marine shells for calibration on an atmospheric curve, many laboratories determine a mean reservoir correction called *R* or *R**, derived from the averaging of R(t) determinations (as presented in Table 5.1). However, whereas series of R(t) values reflect variations in the marine reservoir through time, *R* does not.

## THE NEW ZEALAND MARINE SHELL STANDARD AND CALIBRATION OF NEW ZEALAND ARCHAEOLOGICAL SAMPLES

The most common method employed to derive R(t) for New Zealand has been to use marine shells with a known year of collection. Rafter *et al.* (1972) calculated a marine shell reservoir correction *(R)* of 330 years for New Zealand using this method. This figure, known as the New Zealand Marine Shell Standard (NZMSS), was derived through measurement of the $^{14}$C activity of shell carbonate of modern cockles *(Protothaca crassitesta)* gathered from Pounawea, Otago, in 1955 (Rafter *et al.* 1972:627).

The NZMSS was re-calculated by McFadgen in 1978, by determining the radiocarbon activities of six additional shell samples collected live between 1923 and 1957, from a variety of locations around New Zealand (including Pounawea) (McFadgen 1978b, 1982). This was undertaken to test whether or not the NZMSS applied to all areas of coastal New Zealand (McFadgen 1978b:311). The species used for this analysis were *Protothaca crassitesta, Chione stutchburyi, Paphies ventricosum, Paphies subtriangulatum* and *Dosinia anus.* McFadgen (1978b:313) found the mean radiocarbon activity of his six historical samples (*R*) (after correction for the industrial effect) were slightly different from the value presented by Rafter *et al.* (1972), and proposed that the NZMSS should be further corrected by subtracting 55 years from reported dates (1978:311, 313; 1982:384). (This increased the NZMSS to 385 years, see Table 5.1). Law (1984) revised the Rafter *et al.* (1972) and McFadgen (1978b) calculations and believed that 328 years for the Pounawea standard was a more exact figure. He also reassessed McFadgen's (1978) data, and derived a correction of 59 years, which increased the NZMSS to 387 years.

---

[15] *R* is a culmination of R(t) values, see below.

| Ocean | Reservoir Correction/ Marine Shell Standard | Reference |
|---|---|---|
| Australia | 450 years | Gillespie and Polach 1979, Law 1984 |
| Baltic Sea | 320 years | Mangerud and Gulliksen 1975 |
| Costa Rica | 465 years | Glynn *et al.* 1983 |
| Eastern Jutland | 320 years | Olsson 1983 |
| Southern Ocean | 1300 years | Hiller pers. comm.[16] |
| Denmark | 410 years | Rasmussen pers. comm.[17] |
| France | 400 years | Fortin pers. comm.[18] |
| Vanuatu | 407 years | Burr pers. comm.[19] |
| Tropical | 328 years | Law 1984 |
| Fiji | 412 years | Law 1984 |
| Spitsbergen | 425 years | Olsson 1983 |
| Swedish West Coast | 340 years | Olsson 1983 |
| Norway | 440 years | Mangerud and Gulliksen 1975 |
| Spitsbergen | 510 years | Mangerud and Gulliksen 1975 |
| Arctic Canada | 750 years | Mangerud and Gulliksen 1975 |
| New Zealand | 328 years | Law 1984 |
| New Zealand | 330 years | Rafter *et al.* 1972 |
| New Zealand | 336 years | Jansen 1984 |
| New Zealand | 385 years | McFadgen 1978 |
| New Zealand | 387 years | Law 1984 |
| New Zealand | 430 years | Anderson 1991, IGNS records. |

Table 5.1. Reservoir corrections/marine shell standards from oceanic regions around the world.

This figure altered again in 1984 with the IGNS producing a NZMSS of 336 years (Jansen 1984) (it is not clear whether the IGNS advised researchers to add on a further 55 or 59 years to this figure based on McFadgen (1978) and Law 's (1984) calculations).

The New Zealand example illustrates that although the calculation of an average *R* value was intended to simplify the correction of marine shell CRAs, changes in this value over time has had significant effects on the prehistoric marine shell radiocarbon chronology (see Anderson 1991:782). For example, one *R* value briefly applied to New Zealand marine shell radiocarbon dates between 1955 and 1959 was 430 years (see IGNS sample record form R192/15, and Anderson 1991:782). For New Zealand, six marine shell standards had been derived or refined between 1955 and 1984, three by the IGNS laboratory and three by archaeologists. These changes have brought about confusion in New Zealand archaeology, making it difficult, if not invalid, to compare reservoir corrected dates obtained at differing times. The short duration of New Zealand prehistory, makes the 100 year variation (330 - 430 years) in the NZMSS even more significant.

If archaeologists require an accurate appraisal of the age of a site which has been dated over a number of years by marine shell, then this can only be accomplished by obtaining the original CRAs from the laboratory, and then applying a single reservoir correction before calibration. McFadgen's (1978b) reservoir correction of 385 years can be considered the most applicable for New Zealand at present, due to the number of R(t) calculations employed in deriving his final *R* value. However, as noted above, a fixed *R* does not take into account changes in R(t) through time, if they occurred.

## A NEW MARINE SHELL CALIBRATION METHOD

### The Box Diffusion and Global Carbon Models

Stuiver, Pearson and Braziunas (1986:980) recognised the problems associated with a fixed, country-specific *R* value. The $^{14}C$ concentration of the mixed layer ocean depends on its exchange rate with the atmosphere, and with the $^{14}C$ depleted deep ocean. These exchange rates, though, can change through time because of fluctuations in atmospheric $^{14}C$, or increased storminess. A fixed *R* correction, subtracted from a marine shell CRA to give a reservoir corrected age, could therefore result in erroneous calibrations.

[16] Dr. A. Hiller, Arbeitsgruppe Palaoklimatologie, Quartarzentrum der Universitat Leipzig, Permoserstrabe, Germany.

[17] Prof. K. Rasmussen, Carbon-14 Dating Laboratory, National Museum and Geological Survey of Denmark, Copenhagen, Denmark.

[18] P. Fortin, Centre de Datation par le RadioCarbone, Universite Claude Bernard Lyon, Villeurbanne Cedex France.

[19] Dr. G. Burr, Arizona AMS Facility, The University of Arizona, Tucson Arizona, USA.

Stuiver *et al.* (1986:980) noted that reservoir ages have been determined by radiocarbon age analysis of marine shells of known date of collection, after correction for the industrial affect. However, this method, which uses the equation 1950 - X (where 1950 is the known date of collection of the sample and X is the conventional radiocarbon age of the shell), assumes a constant atmospheric level of $^{14}C$, and that the $^{14}C$ age (X) can be directly subtracted from the calendar date (Stuiver *et al.* 1986:980).

Stuiver *et al.* (1986:980) believed that a second method for calibrating marine shell radiocarbon ages was required, which would take into account changes in the reservoir age of the mixed layer through time. Whereas the original process derived R(t), subtracted this from the marine shell CRA, and calibrated the result on an atmospheric curve (see above), the second method would produce a marine calibration curve that took into account the temporal variability of R(t) (Stuiver *et al.* 1986:980).

A box-diffusion model, based on Oeschger *et al.* (1975), was developed to simulate global carbon exchange (Stuiver *et al.* 1986:980). The Oeschger *et al.* (1975) Box Diffusion Model explains the movement and distribution of $^{14}C$ through the atmosphere, biosphere, mixed layer ocean, and deep ocean reservoirs. It is similar to other box models in that it considers parameters such as the atmospheric $CO_2$ production rate, thickness of the mixed layer ocean reservoir, $CO_2$ concentration in the deep ocean, exchange coefficient between atmospheric and mixed ocean $^{14}C$, and residence times of $^{14}C$ in the various reservoirs (Oeschger *et al.* 1975:171-174, 184). However, Oeschger *et al.* (1975) noted that although box models did explain the long term movement and distribution of $^{14}C$ through the various reservoirs, they were not sensitive enough to illustrate short-term perturbations, such as the influx of fossil fuel or nuclear $^{14}C$ (Oeschger *et al.* 1975:168-169). The mixed layer and deep ocean could be particularly affected by sudden $^{14}C$ influxes, and so specific modelling was required.

Oeschger *et al.* (1975) believed that the deep ocean reservoir could be viewed as a diffusive medium, rather than as a well mixed box (as considered by previous box models). They therefore proposed use of the parameter *K*, which represented a constant vertical eddy diffusivity within the deep ocean reservoir. This value was calculated using measured data on changes in pre-bomb $^{14}C/^{12}C$ ratios with ocean depth for the Atlantic and Pacific oceans (see Oeschger *et al.* 1975:174-176). Oeschger *et al.* (1975:176) also calculated an exchange coefficient between the atmosphere and mixed ocean $^{14}C$ ($k_{am}$) (the calculated value agreeing with those derived in previous box models), to use in association with *K*. Oeschger *et al.* (1975) explained that by incorporating these parameters into their Box Diffusion Model, short term $^{14}C$ perturbations in the mixed layer and deep ocean reservoirs could be effectively simulated.

Oeschger *et al.* (1975) used their Box Diffusion Model to simulate various known short-term $^{14}C$ perturbations in the mixed layer ocean. One example considered the increase in bomb $^{14}C$ in this reservoir between 1956 and 1969 (Oeschger *et al.* 1975:186-188). It was found that their computer model successfully simulated this short-term influx of $^{14}C$, and was in agreement with actual bomb $^{14}C$ increases.

The box diffusion model developed by Stuiver *et al.* (1986:981), called the Global Carbon Model (GCM), used $^{14}C$ atmospheric data for the period AD 1950 to 7746 BC (later, Stuiver and Braziunas (1993) extended the model to 10,000 BC, incorporating coral-based $^{14}C$ data for the period 10,000 - 9400 BC). The Global Carbon Model set parameters for atmospheric $CO_2$ concentration, oceanic C concentration, the amount of C in the biosphere, residence times in the short and long-term reservoirs, gas exchange rate *F* between the atmosphere and mixed layer ocean, and the vertical diffusion coefficient $K_z$ for the deep ocean (see Stuiver *et al.* 1986:981). From this model, a bi-decadal modelled marine calibration curve was constructed for the mixed layer ocean, for the period AD 1950 to 7746 BC (see Stuiver *et al.* 1986: Figures 3 and 5). As the mixed layer modelled marine curve took into account short-term perturbations in atmospheric $^{14}C$, variability in reservoir ages in the mixed layer through time were accounted for. Stuiver *et al.* (1986:981) noted, for example, that during the AD 1950 to 7746 BC period, short-term perturbations have occurred especially during the 200-900 BC period. During this period, a fixed reservoir correction normally used for other time periods may be inapplicable.

The mixed layer calibration curve generated by Stuiver *et al.* (1986, 1993), allows the conversion of a marine shell CRA to years cal AD. The average reservoir correction for the world mixed layer ocean generated by the GCM is 373 years. However, Stuiver *et al.* (1986:982; 1993:138) note that this average correction results from the parameters chosen for their model, and although the model reflects change in R(t) through time for the world mixed layer ocean, it does not take into account regional variations in ocean reservoir ages. This is caused by regional oceanic processes, such as upwelling of $^{14}C$ depleted waters, and when regional atmospheric $^{14}C$ activity differs from globally observed activity (Stuiver *et al.* 1986:982, 1993:138). This constant reservoir difference (ΔR) between the regional and modelled world ocean curve is, (as a first approximation) not time-dependent, and so can be calculated for a region and subtracted from marine shell CRA, before calibration on the mixed layer modelled marine calibration curve.

## Calculation of ΔR

There are two methods by which ΔR may be calculated for use on the Stuiver *et al.* (1986; 1993) modelled marine calibration curves. The first requires the use of pre-1950 shell of known historic age, and the second employs dated terrestrial/marine shell pairs from prehistoric contexts.

The first method was used initially by Stuiver *et al.* in 1986 to produce the first worldwide regional ΔR corrections. For a particular region, a number of marine shell samples of known historic age (AD X), collected prior to 1950, are dated. This radiocarbon age is termed P. The mixed layer ocean $^{14}C$ age, Q, for AD X, is then found using the model-calculated conventional $^{14}C$ age curve for the mixed layer ocean (Figure 5.1). The ΔR for the sample is then calculated by:

$$\Delta R = P - Q$$

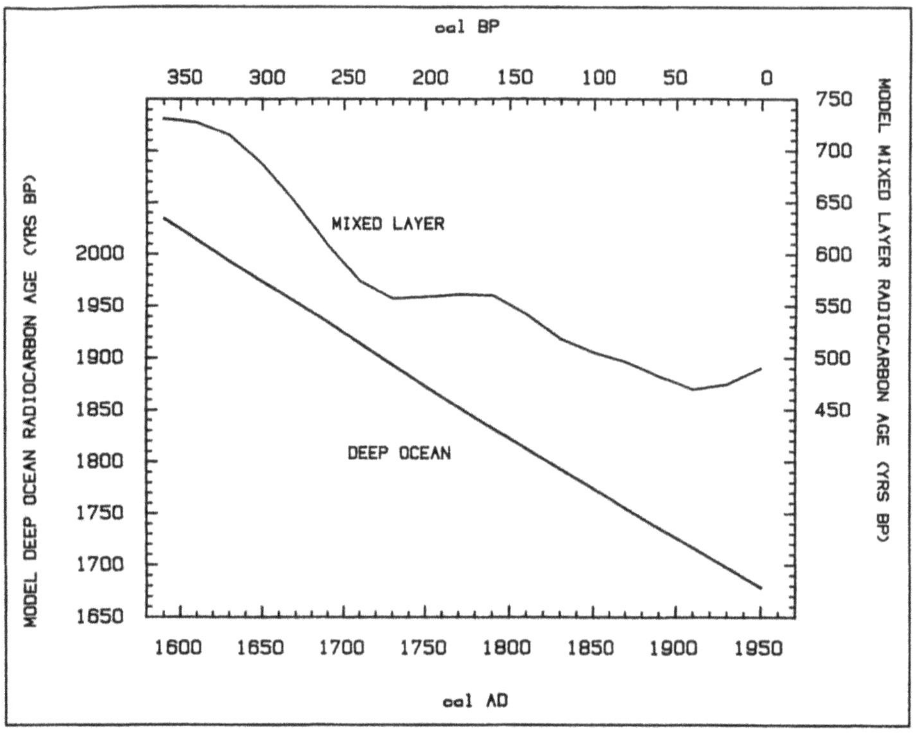


Figure 5.1. Modelled marine conventional radiocarbon age curves for the mixed layer and deep ocean. The mixed layer curve is used to convert a calendar marine shell date (AD X) into a modelled marine age (Q) for ΔR calculation (from Stuiver *et al.* 1986:994).

| REGION | SAMPLE NO. | HISTORICAL AGE (cal AD) | CONVENTIONAL SAMPLE $^{14}$C AGE ($^{14}$C YRS BP) | ΔR ($^{14}$C YRS BP) |
|---|---|---|---|---|
| New Zealand | ---- | 1923 | 416 ± 42 | -57 ± 42 |
| New Zealand | ---- | 1925 | 371 ± 50 | -103 ± 50 |
| New Zealand | ---- | 1949 | 210 ± 41 | -280 ± 41 |
| New Zealand | INS no. R.42 | 1955 | 446 ± 42 | -44 ± 42 |
| WEIGHTED MEAN OF ABOVE 3 SAMPLES (WITH -280 ± 41 EXCLUDED) SCATTER IN UNWEIGHTED MEAN IS 20 YR | | | | -65 ± 25 |

Table 5.2. Original ΔR value for New Zealand calculated by Stuiver *et al.* (1986:1020-1021).

This correction is then applied to marine shell CRA during calibration using the Stuiver *et al.* (1986, 1993) modelled marine calibration curves. Suggested ΔR corrections (presented in 1986) for various regions of the world and calculated using method one, can be seen in figure 5.2. The weighted mean ΔR values for each region were derived from sets of ΔR determined for individual shell samples. The 1986 New Zealand determination can be seen in table 5.2. The 1σ value associated with the ΔR value was derived from the weighted mean of the $^{14}$C ages. However, Stuiver *et al.* (1986:982) note that the 1σ error for the mean ΔR value could also be calculated as an unweighted or 'scatter' mean when the $^{14}$C ages are taken as a data set (Table 5.2). Stuiver *et al.* (1986:982) used the largest of the two sigma determinations for their ΔR corrections, and assigned a regional reservoir correction for New Zealand of -65 ± 25 $^{14}$C years BP (as opposed to -65 ± 20).

The second method for calculating ΔR employs terrestrial/marine shell pairs (Stuiver *et al.* 1993). This method is used to calculate ΔR for the archaeological sites in this study. As we are determining a reservoir correction for the marine environment, the terrestrial (atmospheric) radiocarbon age is first converted to a modelled marine age using a conversion curve (Figure 5.3). This curve plots modelled marine radiocarbon ages against atmospheric radiocarbon ages. The derived modelled marine age is termed the "known historic age" ($Q_{at}$) and is compared to the paired marine shell CRA ($P_{ma}$). ΔR is therefore calculated using the formula:

$$\Delta R = P_{ma} - Q_{at}$$

The ΔR standard error is found using the formula (see Little 1993:464; Higham 1993:126):

$$\Delta R\ \sigma_{total} = \sqrt{(\sigma_{ma}^2 + \sigma_{at}^2)}$$

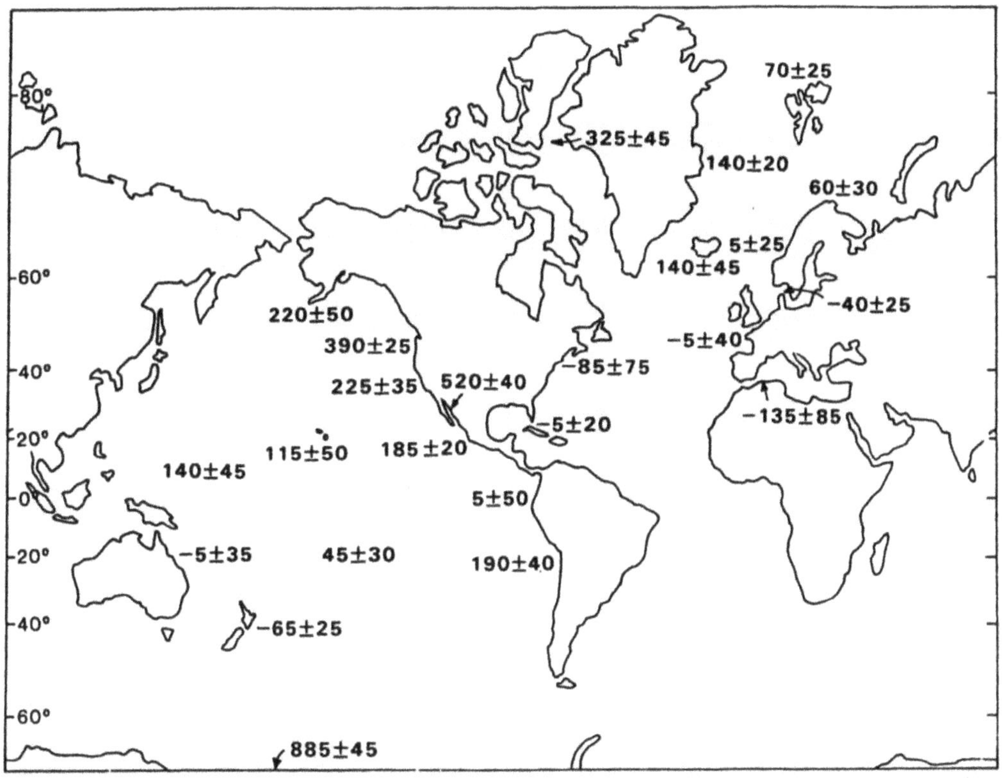


Figure 5.2. ΔR values for oceanic regions of the world calculated by Stuiver *et al.* (1986:995).

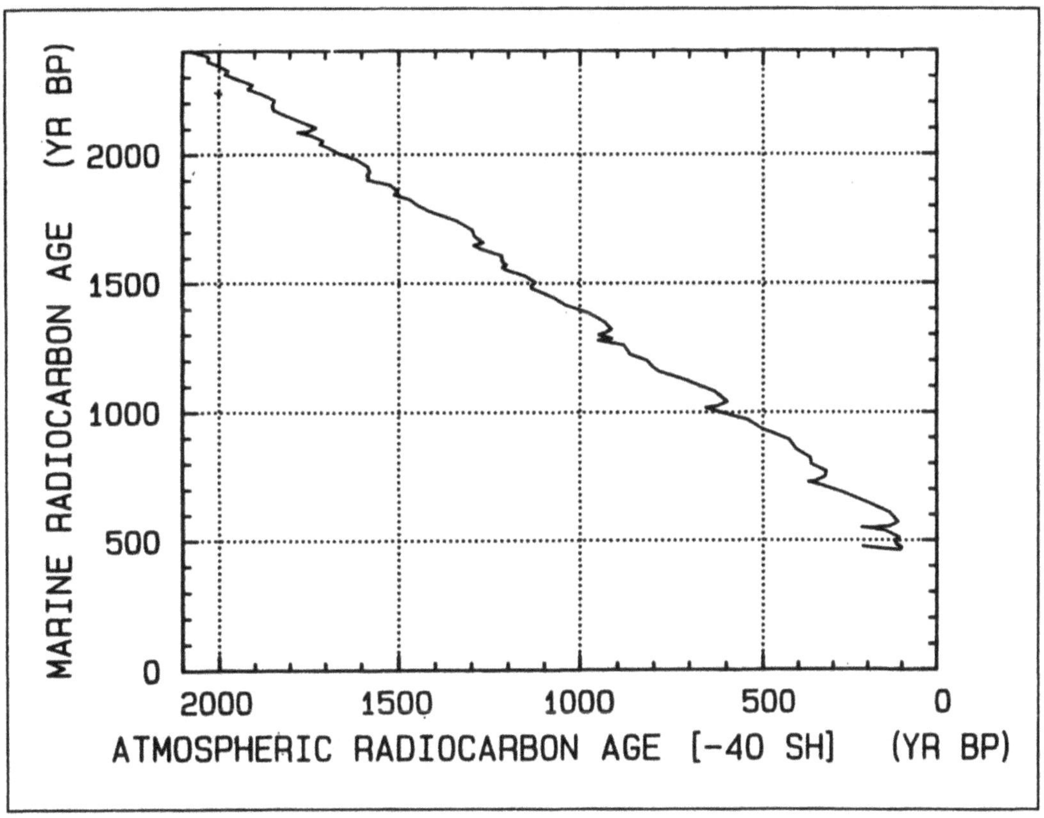


Figure 5.3. Modelled marine CRAs vs. atmospheric CRAs (Northern Hemisphere). This curve is used to convert an atmospheric radiocarbon age, into an equivalent modelled marine radiocarbon age ($Q_{at}$) for ΔR calculation (from Stuiver and Braziunas 1993:154).

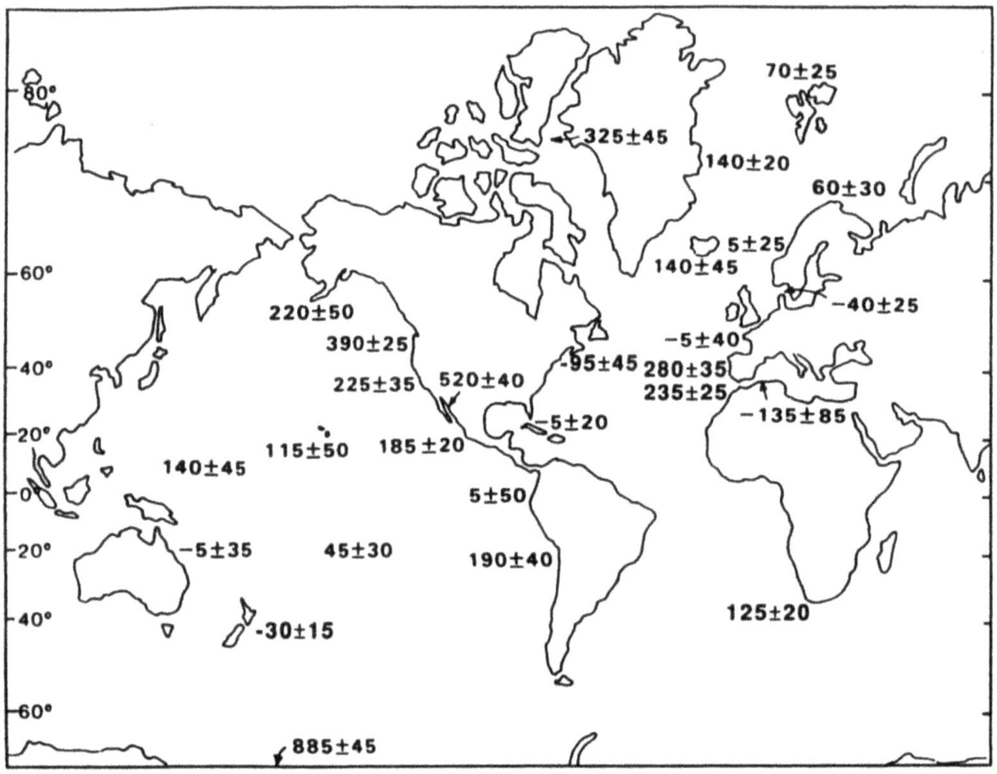


Figure 5.4. ΔR values for oceanic regions of the world (from Stuiver and Braziunas 1993:156).

| Laboratory Response | No. of Laboratories Responded | |
|---|---|---|
| | (n) | (%) |
| Laboratories currently undertaking ΔR research | 9 | 13.4 |
| Laboratories not undertaking ΔR research | 49 | 73.2 |
| Laboratories not undertaking ΔR research but use the Stuiver *et al.* 1986, 1993 value for their region | 9 | 13.4 |
| Total Laboratories Responded | 67 | 100 |

Table 5.3. Current ΔR research around the world.

## REVIEW OF ΔR RESEARCH

Since the proposed regional ΔR values for the world's oceans were calculated by Stuiver *et al.* in 1986, few new ΔR values appeared in the 1993 revision (cf. Figures 5.2 and 5.4) (Stuiver and Braziunas 1993). Of the 24 regional ΔR values calculated in 1986, only two were changed, and three new determinations added in 1993. The ΔR values from New Zealand and the North American east coast altered from -65 ± 25 to -30 ± 15, and -85 ± 75 to -95 ± 45 respectively, and new estimates came from South Africa (125 ± 20) and the coast of Portugal (235 ± 25 and 280 ± 35). Recent studies have calculated ΔR for the eastern Pacific coastal and estuarine waters (ranging between 215 ± 30 and 410 ± 40) (Ingram and Southon 1996:575-576), the Aegean Sea (149 ± 30) (Facorellis *et al.* 1998:969) and the Samoan Archipelago (57 ± 23) (Phelan 1999:23).

Archaeological applications utilising ΔR since 1986, often continue to use values supplied by Stuiver *et al.* (1986), because of the paucity of new data. This is seen in the calibration of marine shell radiocarbon dates from American Samoa (Hunt and Kirch 1987; Kirch, Hunt and Tyler 1989), the Mussau Islands (Kirch and Hunt 1988), and Moloka'i Island, Hawaii (Weisler 1989). For locations where no ΔR value exists, archaeologists have calculated a weighted mean from values for islands in their study vicinity. This was done by Kirch and Hunt (1988) and Kirch, Hunt and Tyler (1989) using values from the islands Eniwetok, Hawaii and the Society Islands to develop a ΔR for the "tropical central Pacific".

I conducted a survey of 181 radiocarbon laboratories throughout the world to determine the extent of ΔR research. I received 67 responses, and only nine of these indicated involvement in this type of research (Table 5.3; Appendix 1). Of these responses, 5 indicated that their laboratory did not date marine shell, and 44 laboratories noted that they still use an *R* correction. Because few laboratories are undertaking this research, the applicability of ΔR for various oceanic regions, and problems associated with its calculation, have not been widely investigated. The literature relating to such research is therefore limited at present. A review of particular studies regarding the calculation and use of ΔR since 1986 is therefore required to determine its merits and short comings for calibrating marine shell radiocarbon ages from New Zealand archaeological sites.

## A Review of ΔR Research Literature

New Zealand appears to lead in research on ΔR. The first study was by McFadgen and Manning (1990), followed by Higham (1993) and then Higham and Hogg (1995). Other studies of note have been conducted by Little (1993) for the north east coast of America, Dye (1994) for Hawaii, and Heier-Nielsen (1995) for the Danish coast. These studies are summarised below.

### *McFadgen and Manning (1990)*

In 1990, McFadgen and Manning (1990:229-232) recalculated the 1986 ΔR correction for New Zealand using 11 radiocarbon ages on marine shell of known historic age. The ΔR values were statistically indistinguishable and ranged from -100 to + 40 years. From these individual values, a weighted mean ΔR correction was derived for New Zealand of -30 ± 13 (McFadgen and Manning 1990:229-230). It was suggested by McFadgen and Manning (1990:230) that when calibrating old marine shell CRAs for New Zealand prehistory using the Stuiver *et al.* (1986) modelled marine curve, the old marine shell age should be converted back to a conventional radiocarbon age by adding on the New Zealand MSS of 336 years. Then, using the CALIB computer program, these dates could then be calibrated using the New Zealand ΔR correction of -30 ± 13 years.

### *Little (1993)*

Little (1993) used 15 paired charcoal and marine shell archaeological samples, and 10 marine shell samples paired with wood of known historic age, to investigate the reservoir effect and local ΔR correction along the north east coast of America. Sixteen locations were sampled from five study areas. The reservoir age R(t) for each terrestrial/marine shell pair from her 16 sampled locations, was found by subtracting the conventional terrestrial $^{14}$C age from its paired conventional marine shell $^{14}$C age. Little (1993) determined ΔR for each location using method one as outlined above, and assigned a mean ΔR for each of the five study areas.

Little found close agreement between the calibrated charcoal and marine shell dates for both prehistoric and historic samples, when using the mean ΔR value from each study area. She did, however, detect anomalous R(t) and ΔR values from some of her sampled locations, and discussed whether the variations in ΔR were either the result of geographic influences, or temporal variability.

Three locations with particularly variable R(t) and ΔR values were Long Island Sound, and the Hudson and Merrimack Rivers. Sampled sites around Long Island Sound, and near the river mouths, had higher ΔR than in the Gulf of Maine. Little proposed that geographic influences may have affected the marine shell ages due to fluvial input at these locations, and marine flushing between the estuaries, Long Island Sound, and the oceanic promontory formed by Cape Cod and its islands (Little 1993:465). The variations in reservoir ages could also have been related to the habitats of the different shellfish species. As shellfish use two sources of carbon for shell building, DIC and metabolic carbon, it was possible that the shellfish with large R(t) values were obtaining $^{14}$C depleted DIC upwelled from Antarctica Intermediate Water. This water contains a pre-bomb and pre-industrial *R* of 760 years, therefore raising the R(t) values significantly (Little 1993: 468).

Little also suggested that the shellfish could have obtained metabolic carbon from sources such as modern or ancient marine and terrestrial plant detritus, including peat or humus, introduced into estuaries by rivers (Little 1993:468). In this scenario, "Dissolved, particulate or metabolised organic carbon from terrestrial sources with low $\Delta^{13}$C (*ca.* -26‰), can reduce shell $\Delta^{13}$C substantially and raise *R* as high as2300 yrs (Keith & Anderson, 1963; Spiker, 1980), or lowering it, depending upon the age of the plant carbon" (Little 1993:468). With these possibilities in mind, Little (1993:469) believed, "A fluvial source for the high ΔR's [from the Hudson and Merrimack Rivers, and Long Island Sound] is supported by an analysis of the reservoir age...".

Little (1993:465, 470) recommended use of Stuiver *et al.'s* (1986) modelled marine curve for the calibration of marine shell CRA, as this method was more accurate than using a fixed *R* correction and atmospheric calibration curve. From her analysis, she calculated a ΔR value of 95 ± 43 $^{14}$C yr for the North East Coast of America. Little (1993:470) suggested that changes in ΔR over time should be investigated, when using prehistoric terrestrial/marine shell pairs, and that for accurate calibration of marine samples, the temporal and geographical characteristics of reservoir ages should be described. She also proposed that from variations in ΔR in her study locations, old terrestrial carbon from humus or peat brought into estuaries by rivers was the probably the main source of the significant discrepancies in reservoir ages in the region (Little 1993:470). The input of Antarctic Intermediate Waters, modern plant detritus, and dissolved limestone were believed to be of secondary importance.

### *Higham (1993)*

Higham (1993) tested McFadgen and Manning's (1990) New Zealand ΔR value using paired charcoal and marine shell radiocarbon dates from the archaeological sites Shag Mouth and Pleasant River in Otago, and from the Awaroa Inlet archaeological site near Nelson.

Higham (1993:125) calculated ΔR using the Stuiver and Braziunas (1993) terrestrial/marine shell pair method (see above and Figure 5.3). For each site, he pooled the statistically indistinguishable charcoal radiocarbon dates, and converted this pooled mean age into a modelled marine age. He then compared this modelled marine age with each marine shell CRA. At the Shag Mouth site, 13 charcoal radiocarbon dates were pooled and then compared to each of the 15 marine shell radiocarbon ages. Higham's (1993:152) calculations derived a ΔR value of 16 for the site with a weighted mean sigma ± 11, and a scatter mean sigma of ±

33. Following Stuiver *et al.*'s (1986) recommendation that the largest 1 standard error should be used of the weighted and unweighted ("scatter") means, Higham found that his ΔR value of 16 ± 33 was statistically indistinguishable from McFadgen and Manning's (1990) ΔR value of -30 ± 13 (Higham 1993:128-129).

At the Pleasant River site, ΔR values varied greatly because of large age discrepancies between different shellfish species. Initially Higham (1993:152) believed that the age differences were possibly due to species habitat and environmental influences. Later analysis of the site found that a sand deposit, which once separated two cultural layers, had gradually been eroded over time by tidal action. This process therefore mixed cultural material from two past and unrelated events (Higham pers. comm.). At the Awaroa Inlet site, sample provenance and post-depositional history, were secure (as at Shag Mouth). A ΔR value of -55 ± 39 was derived from this site using two charcoal and two marine shell radiocarbon dates. This ΔR value was also statistically indistinguishable from the McFadgen and Manning (1990) correction.

***Dye (1994)***

Dye (1994) investigated ΔR for various coastal areas in Hawaii, using both marine shells of known historic age, and charcoal/marine shell pairs. This study also examined the effect of coastal volcanic and limestone geology on the radiocarbon dating of shellfish species.

Dye observed that ΔR calculated from the eight marine shells of known historic age, showed significant variation. Although the shells dated were collected between 1918 and 1947, the ΔR values ranged from -500 to 3820 years. He suggested the large ΔR value of 3820 years from Pearl Harbour was possibly due to pollution of the harbour by fossil fuels during U.S. naval occupation (Dye 1994:51). For the remaining seven ΔR values, Dye (1994:52) noted, "No hypotheses have been advanced to account for the extreme variability in the apparent ages of these shells.".

Dye (1994:52-54) then compared 11 paired charcoal and marine shell radiocarbon ages from the archaeological site H8 at Wai'ahukini in Hawaii, and from sites at Barbers Point, O'ahu. He observed significant age differences between seven of the paired charcoal and marine shell CRAs from the sites. This in turn led to large discrepancies in ΔR values between sites. Although no explanations were proposed for the age differences observed at site H8 in Hawaii (other than possible problems with the radiocarbon dates on wood charcoal for two pairs), Dye (1994:52) did express his concerns regarding the provenance of the samples, and the shellfish species were unknown. Dye did not specify possible reasons for the extreme age differences between the charcoal and marine shell pairs from the Barbers Point sites, or the large ΔR difference between this site and site H8, but he did note that Barbers Point is located on the 'Ewa Coral Plain (Dye 1994:54).

Dye (1994:55) examined the effect of dating marine shells from two different coastal geological environments, one consisting of Pleistocene limestone, the other formed from Holocene lava, using shellfish species commonly found in Hawaiian archaeological sites. The marine shells selected for dating originated from locations with similar geology as found at site H8 (Holocene lava) and the Barbers Point sites (Pleistocene limestone). Marine shell from the limestone environment showed great $^{14}$C ages and higher ΔR values than the same species dated from the volcanic environment (Table 5.4).

The ages differences from the shells dated from the limestone coast were explained as being due to two factors: 1) the environment in which the shellfish developed ,and 2) the feeding habits of particular shellfish species. Dye suggested the older ages may have resulted from the Pleistocene limestone being a source of old carbon, which was ingested by the gastropods either from the algae fed upon, or directly from the limestone they scraped while feeding (Dye 1994:55). Dye (1994:55-56) believed that the age of the gastropod sample Beta-54331 being younger than the other two gastropod samples (Beta-54332, Beta-54333) from the intertidal and splash-zone limestone area, may be due to its shallow offshore habitat, or difference in diet (which also included Holocene reef and reef detritus). Factors other than geology identified as possibly influencing the dates on shell from the volcanic coastlines, were upwelling and riverine carbon. Differences in carbon uptake between shellfish species was also seen as a possible influence on the radiocarbon age of marine shells from Hawaii.

| **Lab no.** | **Species** | **Substrate** | **Colln. yr** | **Convent. $^{14}$C age** | **ΔR value** |
|---|---|---|---|---|---|
| Beta-54332 | *C. exarata* | Limestone | 1914 | 1270 ± 80 | 800 |
| Beta-54336 | *C. exarata* | Volcanic | 1923 | 740 ± 270 | 270 |
| Beta-54331 | *C.caputserpents* | Limestone | 1915 | 980 ± 80 | 510 |
| Beta-54334 | *C.caputserpents* | Volcanic | 1923 | 730 ± 80 | 260 |
| Beta-54333 | *N. picea* | Limestone | 1930 | 1230 ± 80 | 760 |
| Beta-54335 | *N. picea* | Volcanic | 1924 | 610 ± 80 | 140 |

Table 5.4. "AMS Dates on Marine Shells from Pleistocene Limestone and Holocene Volcanic Coast." (from Dye 1994:55).

Dye (1994:56) concluded that variation in dates on marine shell from Hawaii could possibly introduce uncertainty in calibrating samples from archaeological sites, but identification of factors which influence the dating of shells could aid in reducing errors in ΔR values. He believed that different ΔR values for limestone and volcanic coastal regions, and for specific shellfish species, should be useful in particular archaeological situations (Dye 1994:56).

### *Higham and Hogg (1995)*

Higham and Hogg (1995) determined a new ΔR value for New Zealand using the McFadgen and Manning (1990) historic shell data, fish otoliths of known historic age from Kalish (1993), and the Shag Mouth archaeological site paired charcoal and marine shell radiocarbon dates. This paper is particularly significant for world ΔR research, because not only are both methodologies employed to determine this local reservoir correction, but it is the only published research where all sample type factors have been fully addressed. The historically collected samples had careful provenance and the prehistoric charcoal used for calculating ΔR had been identified to twigs of short-lived tree and shrub species.

The Kalish (1993) snapper (*Pagrus auratus*) otoliths were collected between 1918 and 1955 from the east coast of the North Island, and were originally used by the researcher to investigate pre- and post-bomb radiocarbon activity in the southern temperate ocean. Higham and Hogg (1995) calculated ΔR values from the seven AMS ages taken on the otoliths, and from McFadgen and Manning's (1990) earlier historic shell data. The Kalish (1993) ΔR values were found to be statistically indistinguishable from McFadgen and Manning's (1990) ΔR value of -30 ± 13, producing an overall error-weighted mean ΔR correction of -25 ± 15 $^{14}$C years (Higham and Hogg 1995:411).

Paired charcoal/marine shell dates from Shag Mouth (see above), were then used to test this value (Higham and Hogg 1995:411-415). The revised calculated ΔR for this site is 13 ± 17 (due to the addition of four shell ages) (Higham and Hogg 1995:415). This value was observed as statistically indistinguishable from the Higham and Hogg (1995) ΔR estimate of -25 ± 15, but statistically different from the McFadgen and Manning (1990) ΔR of -30 ± 13. Higham and Hogg (1995) then utilised the -25 ± 15 ΔR to calibrate the pooled mean marine shell CRAs from Shag Mouth, for comparison with the pooled mean calibrated charcoal date. The calibrated marine and charcoal radiocarbon dates matched comfortably (Higham and Hogg 1995:415).

Higham and Hogg (1995:415) concluded that the new ΔR value would accurately calibrate marine shell radiocarbon ages from Shag Mouth, and that the regional ocean appears to have been stable in the recent past. They believed that a knowledge of the local environment from which terrestrial/marine shell pairs are derived was vital in calculating ΔR, to account for factors such as the hard water effect. Also, the identification of charcoal to twigs from short-lived flora species prior to dating was critical. Higham and Hogg (1995:415) noted that as New Zealand's ΔR value may change, it was the responsibility of the researcher who has obtained the marine shell dates to publish the CRAs, and identify the ΔR value and calibration curve used (Higham and Hogg 1995: 415).

### *Heier-Nielsen et al. (1995)*

Heier-Nielsen *et al.* (1995) measured 28 AMS $^{14}$C dates from molluscs collected live between 1885 AD and 1945 AD from Danish waters. Fourteen of these samples came from the Danish coast at Skagerrak-Kattegat and from the Belts, with the other 14 coming from Danish fjords. From these measurements, reservoir ages (R(t)), and ΔR were calculated for the various locations. It was found that the ΔR value for coastal Denmark was 13 ± 16 yrs (having a scatter standard error of ± 21 yrs), which was consistent with an earlier calculation of -33 ± 27 yrs for the Norwegian-Swedish Skagerrak-Kattegat coasts. These results also agreed with ΔR values from the North Atlantic and the North Sea of 0, and the United Kingdom of 1 ± 8. However, the remaining 14 shell ages from the Danish fjords gave consistently high R(t) and ΔR values. It was found that fresh water run-off and ground water flowing into the fiords contained dissolved soil carbonate low in $^{14}$C, therefore increasing the apparent shellfish age. It was concluded by Hier-Nielsen *et al.* (1995), that although the Danish, Norwegian and Swedish coasts were part of a uniform $^{14}$C reservoir, the fiords were not uniform because of the influx of fresh water.

## DISCUSSION AND CONCLUSIONS

Early observations by radiocarbon laboratories on differences in radiocarbon age between terrestrial and marine samples, led to the development of country specific Reservoir Corrections (*R*, *R**) or Marine Shell Standards (MSS). These corrections were applied to marine shell radiocarbon ages for conversion to terrestrial dates, so they could be calibrated using atmospheric calibration curves, where the relationship between radiocarbon ages and calendar years was known. This method for calibrating marine shell CRAs is still the most common method used by radiocarbon labs and researchers today. The *R* values used are based on an average of R(t) calculations, which reflect variations in apparent age between marine shell radiocarbon dates and known year of collection of the samples, or paired terrestrial $^{14}$C dates. *R* values have progressively changed in some countries (such as New Zealand), as more R(t) data has been collected. This has brought about confusion for archaeologists when comparing marine shell radiocarbon ages from sites which have been dated over a number of years.

Stuiver *et al.* (1986) recognised the temporally fluctuating environment of the mixed layer ocean, and developed a modelled marine calibration curve based on a Global Carbon Model. This calibration curve provided a standard method based on ocean $^{14}$C reservoir parameters by which marine shell CRAs could be calibrated, and avoided the use of the atmospheric curve, which the reservoir correction method required. The modelled marine curve considered short-term fluctuations of $^{14}$C in the mixed layer ocean, thereby correcting the deficiencies of using a fixed *R* or MSS. Stuiver *et al.* (1986) also introduced a regional marine reservoir correction, termed ΔR, which took into account regional ocean marine reservoir differences, and which was time independent.

However, problems have arisen in the application of Stuiver *et al.*'s (1986, 1993) new marine shell calibration method. These must be addressed if we are to evaluate its applicability for calibrating marine shell radiocarbon ages from New Zealand archaeological sites.

These problems stem from three observations. Firstly, the number of marine shell samples from each oceanic region used in the 1986 ΔR calculations was small, and few additions to the original calculations have been made since this time. Secondly, problems which can affect the calculation of local ΔR, highlighted in subsequent studies, were not considered for the samples used in determining the 1986 values. Thirdly, the methodologies employed for deriving ΔR for various regions by researchers since 1986 have been inconsistent.

Table 1 supplied by Stuiver *et al.* (1986:1017-1021), shows the number of marine shell samples used to derive regional ΔR around the world averages 5, with a range of 2 to 9 samples. The number of samples was therefore small, when compared to the number employed by McFadgen and Manning (1990), Talma (1990), Higham (1993), Little (1993), Dye (1994), Higham and Hogg (1995) and Heier-Nielsen *et al.* (1995). It was found for New Zealand, that with an increase in the number of samples employed to derive ΔR, the original value changed dramatically. This same problem was encountered with the calculation of the New Zealand Marine Shell Standard. Also, because little updating or addition to the original 1986 calculations has occurred, archaeologists calibrating marine shell radiocarbon dates from various sites in Polynesia, for example, still use the 1986 values, some averaging sets of these ΔR corrections from areas close to their location of study.

Geographic and environmental problems have been shown to affect ΔR. Anomalous marine shell radiocarbon ages result from the hardwater effect, upwelling, the use of either open beach or estuarine species, and the intake of metabolic carbon by particular shellfish species. As it is unclear whether these factors were considered in the calculation of the original ΔR values, the Stuiver *et al.* (1986) ΔR values should be viewed as first approximations only.

Studies where geographic or environmental factors have been proposed to be responsible for anomalous ΔR values, should also be closely scrutinised as methodological problems may be the reason for the unusual values. For example, extreme variations in ΔR values observed by Little (1993) and Dye (1994), possibly resulted from using old wood in their terrestrial/marine shell age comparisons. The wood used as the terrestrial comparison in these studies was not identified to twigs of short-lived species (cf. Higham and Hogg 1995). In addition to this, the historic shell used by Little (1993) had no specific provenance data, nor had the terrestrial/marine shell pairs from Dye's (1994) Hawaiian archaeological site H8.

Although Stuiver *et al.* (1986, 1993) set down clear guidelines for calculating and reporting ΔR, some researchers have not followed these recommendations. Sometimes incorrect calculations have been used, and at other times the reported values are too precise (see Higham 1993:128-129). For example, though the McFadgen and Manning (1990:231) ΔR value of -30 ± 13 $^{14}$C years for New Zealand was correct, the largest sigma was incorrectly reported as ± 41 $^{14}$C years, causing confusion for some archaeologists as to which sigma should be used. Of the studies reviewed, only Higham (1993), Higham and Hogg (1995), and Heir-Nielsen *et al.* (1995), strictly followed the method of Stuiver *et al.* (1986, 1993). These studies illustrate the need for more regional ΔR measurements.

An additional problem is the inconsistent advice given to archaeologists on how to calibrate their marine shell CRAs. An example of this is McFadgen and Manning's (1990) recommendations to archaeologists when using Stuiver *et al.'s* (1986, 1993) modelled marine calibration curve. They suggested that researchers should add the NZMSS of 336 years to their old reservoir corrected marine shell dates to convert these back into a conventional radiocarbon age. However, as the marine shell standard has varied by 59 years since 1972, the application of 336 years cannot be applied to all old reservoir corrected marine shell radiocarbon dates. This would not only produce a series of false CRAs, but would also provide inaccurate calibrations on the new modelled marine calibration curve.

In conclusion, the introduction of a modelled marine calibration curve has greatly simplified the earlier difficulties experienced by researchers when calibrating their radiocarbon dates using changing *R* corrections or marine shell standards. It has also meant that the method for calibrating marine shell radiocarbon dates should be globally consistent. However, researchers using or deriving ΔR should be careful not to repeat the mistakes which have occurred with use of the reservoir correction method. It is vital that when researchers publish their radiocarbon data, they provide the CRAs, inform the reader which ΔR value they employ, and the calibration curves utilised (see Higham and Hogg 1995:415). This review has found that few studies have been consistent in the methodology used to derive ΔR, recommended by Stuiver *et al.* (1986, 1993). Additionally, errors made in calculating *R,* were also made in calculating ΔR. Examples include: not identifying charcoal; not considering the habitats and lifeways of shellfish and plant species; lack of data concerning sample provenance; and not considering the post depositional history of the site being dated. All these factors must be considered if researchers are to test the reliability of various marine shell species for dating prehistoric contexts.

Chapters 4 and 5 have identified the potential problems when dating charcoal and marine shell samples from archaeological sites, and when comparing calibrated charcoal and marine shell CRAs. In the following chapter, descriptions of the archaeological sites sampled for this dissertation to determine the reliability of marine shell for radiocarbon dating New Zealand prehistory are presented.

# CHAPTER 6

## ARCHAEOLOGICAL EXCAVATIONS

*"First, it was clear to me that the next advance in our knowledge of human achievement outside the historical field was ependent upon fresh and methodical discovery, and that fresh discovery in great measure meant fresh digging." Sir Mortimer Wheeler (1955:54).*

### TRODUCTION

is chapter describes the 11 archaeological sites excavated d sampled for this study. The aim of these excavations was obtain charcoal/marine shell pairs to investigate the use of arine shell for radiocarbon dating New Zealand prehistory. nese sites were selected to answer particular questions garding the dating of marine shell from various regions of e country.

he Kokohuia archaeological site was selected to investigate e possible influences of the hard water effect on shell ages om a variety of shellfish species living in a calcareous vironment (Figure 6.1). As limestone strata are extensive New Zealand, the hard water effect is probably more of a ncern to archaeologists than upwelling. In contrast to this vironment, archaeological sites at Aotea, Ponui Island, igar Bay and Sawpit Point Inlet were sampled, as these tes are free of calcareous rock. In dating the sites using oth charcoal and marine shell, issues in radiocarbon dating ew Zealand prehistory, such as inbuilt age of charcoal, ould also be addressed.

etailed descriptions of the geology surrounding the sites d environmental conditions during their formation, are esented. This is important to explain any anomalous shell d charcoal ages which may be due to the environmental nditions surrounding the sites in prehistory.

ample provenance is a vital factor when considering the covery of samples from archaeological sites. This entails ot only confirming that the charcoal/marine shell pairs are om exactly the same location, but also that the cultural yer sampled has been undisturbed since deposition and presents a single human-induced event. Sites for this study ere chosen with this criteria in mind.

ll archaeological sites on Ponui Island and Ligar Bay were xcavated and sampled by a team which included the author. he Kokohuia site was visited after completion of the xcavation, and samples selected for dating by the excavator nd the author. The two sites on Aotea and the Sawpit Point te, were sampled by archaeologists involved in the xcavations.

### OKOHUIA (O06/317), OPONONI

#### rchaeological Excavation and Analysis

okohuia is located 4 km from the mouth of the Hokianga arbour near Opononi, Northland (Figures 6.1 and 6.2). The te is considered an early settlement in Maori tradition. Its cation provided a route to resources in the hinterland of Northland both in prehistoric and historic times (Taylor 1995:3). The site sits upon a ridge known as Waiarohia Point, on the eastern side of the harbour, looking across to the massive sand dune complex on the opposite side of the water (Taylor 1995:5) (Figure 6.3). The ridge itself once extended to the sea and contained a number of archaeological features, but many were destroyed when the end of Waiarohia Point was removed to make way for State Highway 12 (Taylor 1995:5). On the surviving section of ridge, few prehistoric features now remain due to the building of roads and houses on the site in the early 1960s and 1989 (Taylor 1995:5).

Taylor excavated five areas in 1991 on the remaining undisturbed portion of the ridge. These excavations showed that the ridge-top acted as a *kainga* in prehistory, where activities such as fishing, food preparation, cooking, adze making and gardening took place. The Area 3 excavation was of particular importance in determining a chronology for the site. This location constituted a large mound on the western edge of the ridge measuring 10 x 6 m in area, and was excavated as two squares, the Upper Midden and the Lower Midden (Taylor 1995:8). The Lower Midden revealed four distinctive layers of occupation, and contained a wide variety of shellfish species and copious amounts of charcoal (Figure 6.4).

A radiocarbon date (Wk-1086) already obtained by Taylor on *Paphies australis* from Layer 4 of this midden gave a calibrated age of 1411 - 1477 cal AD[20]. The Lower Midden was sampled for this study because of the varieties of mollusc and large quantity of charcoal present, as well as an already existing radiocarbon chronology. Material for dating was obtained from Layers 2 to 5.

Each layer of the Lower Midden contained extensive amounts of shell, hangi stones, bone and charcoal, but although similar in content, the layers differed in matrix. Layer 2 was "black and sticky", its constituents being more broken and mixed than the lower layers. Midden in Layer 3 was mixed with a clean brown sand, which possibly originated from the top of the ridge, and was used to cover the midden in prehistory for occupation. The Layer 4 midden contained little sand, and was covered by a distinctive clean coarse yellow sand which sealed it from Layer 3. Layer 5 comprised of midden mixed with a clean yellow sand which originated from people living upon the natural ground

---

[20] Shell CRA calibrated using the modelled marine curve (Stuiver and Braziunas 1993) and New Zealand $\Delta R$ value of $-25 \pm 15$ (Higham and Hogg 1995).

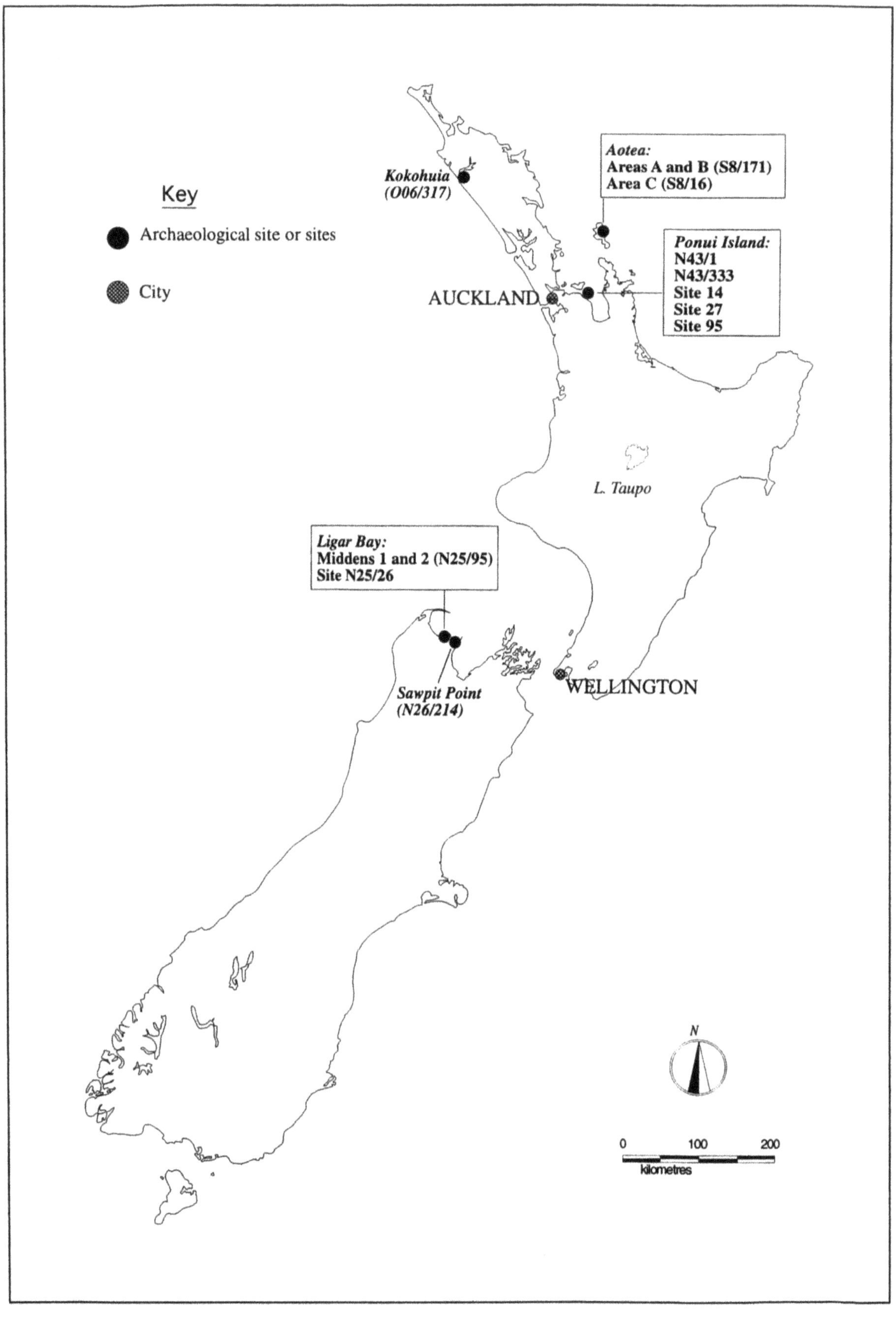


Figure 6.1. Location of archaeological sites sampled for charcoal/marine shell pairs in this study.

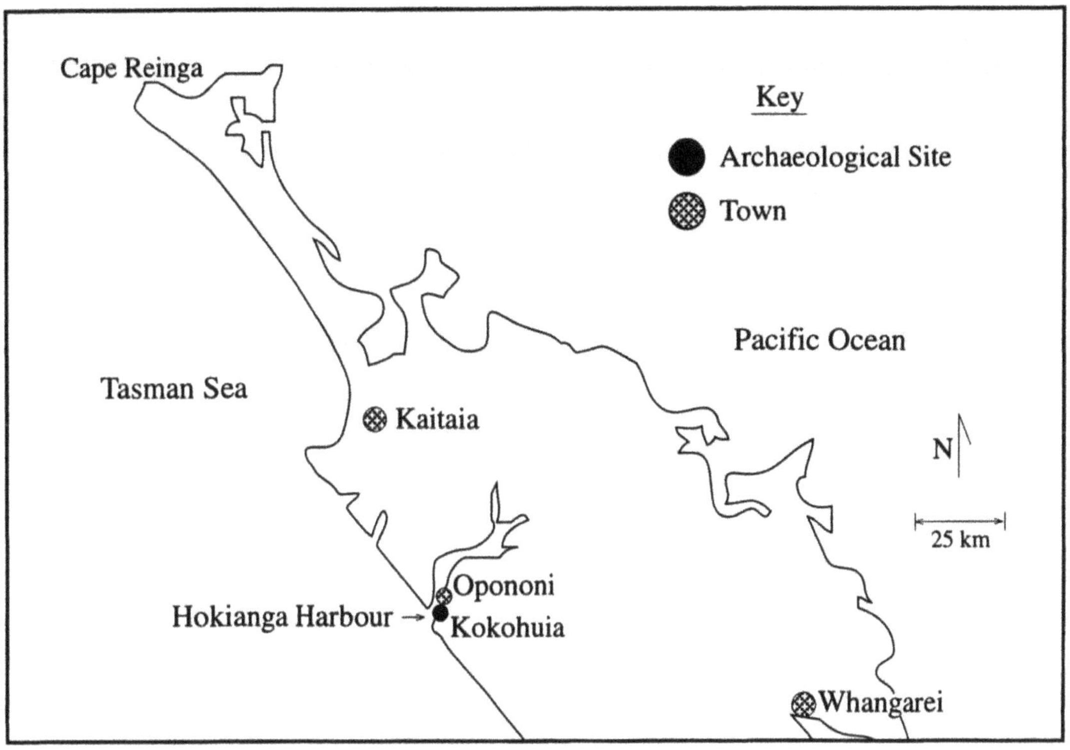


Figure 6.2. Location of Kokohuia in the Hokianga Harbour, Northland.

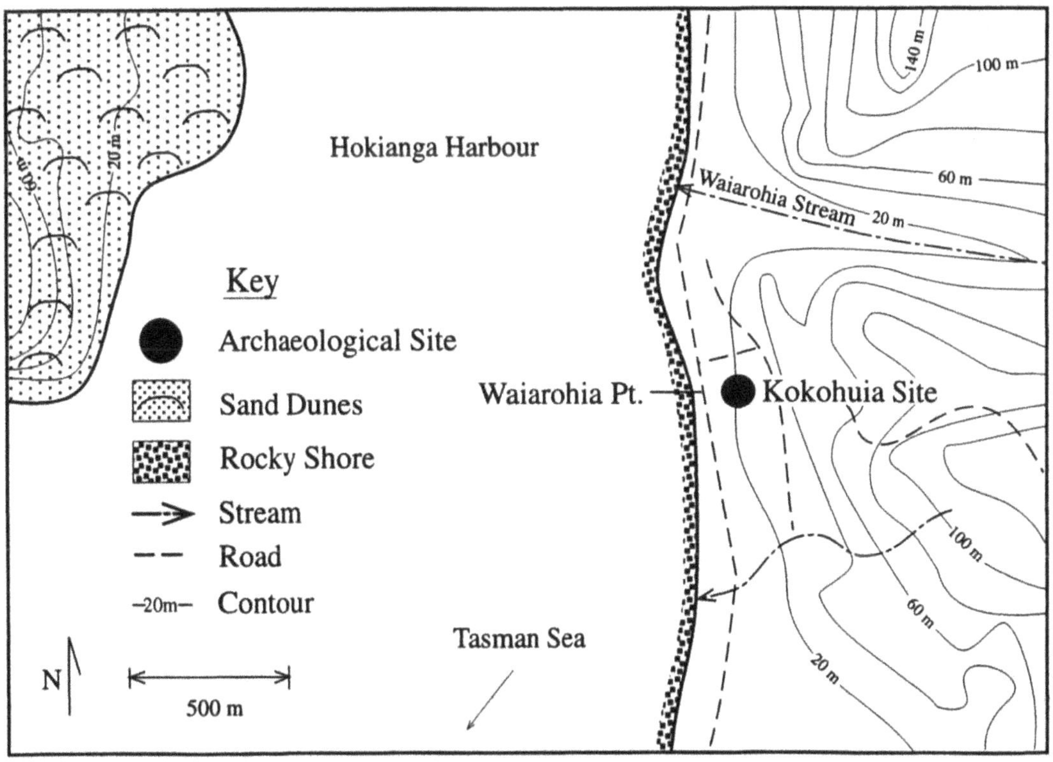


Figure 6.3. Location of Kokohuia archaeological site (O06/317)in the Hokianga Harbour.

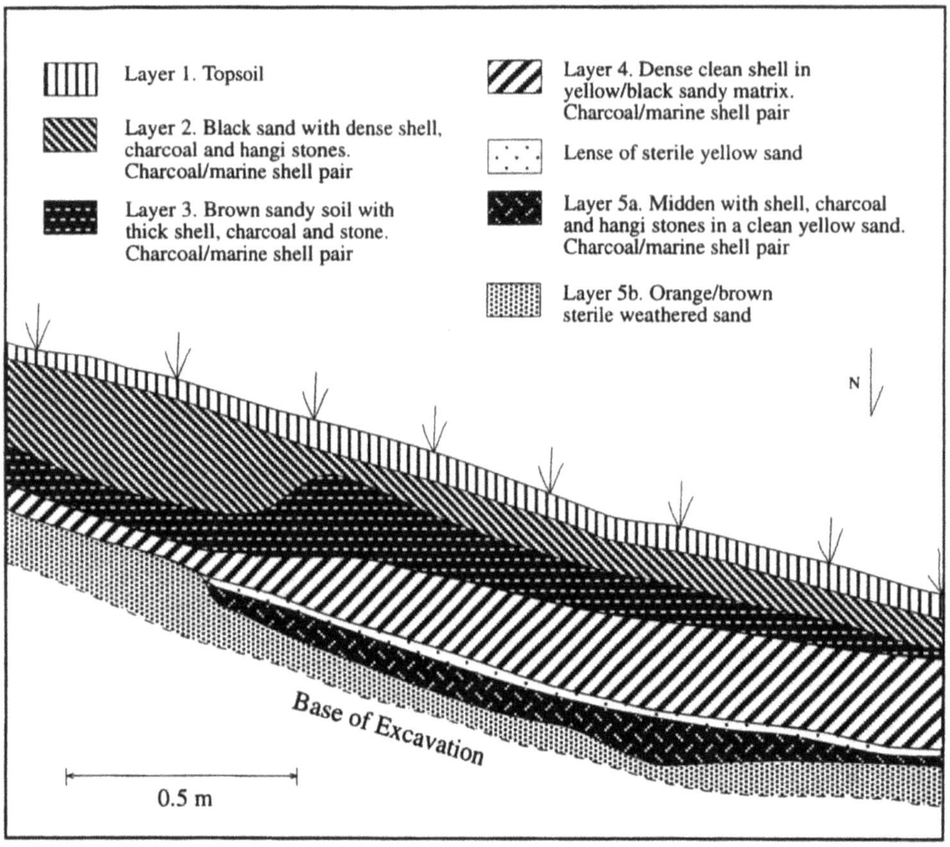


Figure 6.4. Stratigraphy of Kokohuia (O06/317).

surface and churning the sand up with their activities (Taylor 1995:9).

Taylor noted that each midden layer had pockets of single species of shellfish which had been exploited and consumed. The shell from the layers also looked fresh, and retained their original colouring (Taylor 1995:9). Typical features found in the cultural layers were fires, especially in Layer 5 which contained a number of overlapping fire scoops, as well as associated fishhooks. Artefacts recovered during the excavations included 31 pieces of shell fishhooks, a section of bone lure and a full shell lure, obsidian flakes from non-local sources, a bevel from an adze in Layer 2, two small limestone adzes (one from Layer 3 the other from Layer 4), and a Duff 1A black argillite adze found in Layer 5 (Taylor 1995:10,16).

Shellfish species present in all layers included *Paphies australis*, *Austrovenus stutchburyi*, *Venerupis largillierti*, *Lunella smaragda*, *Cominella virgata*, *Perna canaliculus*, *Maoricolpus roseus*, *Cellana* sp., *Melarapha olizeri*, *Amphibola crenata* and *Haliotis iris*. Fish species exploited were as varied as the shellfish. *Pagrus auratus* (Snapper) was the most common species taken, equal in proportion to *Aldrichetta forsteri* (Yellow-eyed Mullet) and *Mugil cephalus* (Grey Mullet) combined. *Nemadactylus macropterus* (Tarakihi), *Arripis trutta* (Kahawai), *Pseudocaranx dentex* (Trevally), *Trachurus declivus* (Jack Mackerel) and *Labridae* sp. were the other main species identified. Also found in the midden were *Chelidonichthys kumu* (Red Gurnard), *Parapercis colias* (Blue Cod), *Thyrsites atun* (Barracouta), *Seriola lalandi* (Kingfish), *Latris lineata* (Trumpeter fish) and shark (unknown species) (Taylor 1995;11). Bird and mammal bone identified from the site indicated the exploitation of *Nestor meridionalis* (Kaka), *Callaeas cinerea* (Kokako), *Rattus exulans* (Kiore) and *Canis familiaris* (Kuri) throughout the occupation of the site, as well as possibly moa (unknown species) during activities in Layer 5.

From the archaeological evidence, Taylor interpreted Area 3 as a food preparation, cooking and consumption area, and even though the cultural events were distinctive, they possibly occurred within a few years (Taylor 1995:12). Activity on this area of the ridge was believed to have occurred in the late Archaic, based on the fish hook and adze typologies, and the single radiocarbon date (Taylor 1995:17). Further radiocarbon dating would confirm the beginning and duration of occupation on the site.

## Environmental Reconstruction

During its occupation, Kokohuia was surrounded by coastal and lowland kauri-podocarp-hardwood forest containing the large trees *Agathis australis* (Kauri), *Beilschmiedia tarairi* (Tarairi), *Beilschmiedia tawa* (Tawa), *Vitex lucens* (Puriri), *Elaeocarpus dentatus* (Hinau) and *Prumnopitys taxifolia* (Matai). Associated with these large trees were smaller trees and shrubs typical of this habitat, such as *Plagianthus regius* (Manatu), *Melicytus ramiflorus* (Mahoe) and *Pseudopanax crassifolius* (Horoeka) filling the mid-canopy up to 15 metres. The lower canopy and forest margins consisted of *Leptospermum scoparium* (Manuka), *Kunzea ericoides* (Kanuka), *Dodonaea viscosa* (Akeake), *Myrsine australis* (Mapau), *Pseudopanax arboreus* (Puahou), *Olearia* species and *Mida salicifolia* (Maire). Particular to the coast and directly associated with the site were *Metrosideros excelsa* (Pohutukawa), *Hebe* species and *Avicennia marina var. resinifera* (Manawa), possibly from along the shore close to the site. Seeds found through all levels of the site, but mainly in Layer 4, included *Vitex lucens, Elaeocarpus dentatus, Podocarpus ferrugineus* (Miro) and *Beilschmiedia tawa,* again showing a close association of the site with the forest. This reconstruction of the prehistoric vegetation at Kokohuia based on charcoal identifications, is also supported by early historic records of the Hokianga harbour where a forested landscape was noted for its flax and kauri resources (Taylor 1995:2). This landscape stands in contrast to today's scrub-coated hillsides.

As noted above, a wide variety of shellfish species were identified from Kokohuia. The habitats of these species suggests that Kokohuia was surrounded by a very different marine environment in prehistory than observed today. All rocky and sandy shore shellfish species were available to the prehistoric occupants (Taylor 1995:2). The reduced shellfish variety today is probably due to hillside erosion from forest clearance and roading activities (Taylor 1995:2). Fish species caught today are similar in variety to those caught in prehistory with snapper, kahawai, mullet, and cod still available. Strong tidal currents in the harbour result in inlets and bays within the harbour being well flushed. However, it is possible that this scenario was different in prehistory, because of the lack of sandy shore shellfish, and dominance of rocky shore shellfish species today.

The Kokohuia site provides an interesting study into the effect of a calcareous geology on the dating of an archaeological site. Kokohuia rests upon the ridge of a Pleistocene consolidated ancient sand dune of Hawera age, overlain by Holocene dune sand deposits (Milligan 1959: Map 1a; Kear and Hay 1961; Taylor 1995:1-2). The Motatau Group formation of the Lower Oligocene lies beneath these dune deposits, and contains argillaceous and crystalline limestone with glauconitic sandstone beds (Kear and Hay 1961; Thompson 1978:444; Taylor 1995:1-2). These limestone and sandstone beds extend out into the harbour to the west of the site, creating a rocky shore covered by the shifting sands of the harbour. To the east of the site and inland, lie beds of calcareous sandstones and siltstones as well as the massive Upper Cretaceous Tangihua Volcanics of marine origin (Challis 1978:628). Extending 25 km up the length of the harbour, the dominant rock formation found exposed from the edge of the harbour and inland creating a buffer of not less than 5 km wide on either side, are sediments of the Mata Series formed during the Late Mesozoic (Hay 1978:368). This series consists of hardened dark siliceous shales, overlain by concretionary micaceous sandstone with carbonaceous bands, containing floral and faunal fossil material from mainly the Haumurian Stage, with a minor Piripauan Stage presence (Kear and Hay 1961; Stevens 1978:362-366; Hay 1978:367-368).

Other than the massive recent dune formations on the western side of the harbour and the more ancient dunes upon which the site itself rests, the geology surrounding the Kokohuia site consists of sediments with a high calcareous content. There is therefore the possibility of a hardwater effect upon shellfish, not only from within the harbour itself, but from the fluvial input, such as the nearby Waiarohia stream, whose catchment includes calcareous Tertiary rock.

## AOTEA

### Archaeological Excavations and Analysis

The large island of Aotea, or Great Barrier Island, is located 20 km north of the Coromandel Peninsula (Figure 6.1 and 6.5). Excavations were undertaken in 1994 by the University of Auckland Anthropology Department, at five locations at Whangapoua on the north eastern shore of Aotea. The aims of the excavations were to teach archaeology students excavation and mapping techniques, and to compare the dating methods of obsidian hydration and radiocarbon dating.

Three areas (A, B and C) from the five excavations were chosen for sampling to investigate the dating of marine shell on Aotea (Figure 6.6). Areas A and B of site S8/171 were located on a natural terrace next to a creek which flows into the Whangapoua estuary. Area C of site S8/16 was located 600 m to the east of A and B on an artificial terrace of a *pa* site. The age of Areas A and B were estimated at *ca.* 250 years BP, based on deposit depth and obsidian hydration dates. Area C was thought to be <300 years old, based on deposit depth, obsidian hydration dates, and because it was a *pa* site, construction would have been after 1500 AD (Figure 4 from McGhie 1994:25) (see Schmidt 1993:58;1996a).

The Area A excavation measured 4 x 8 m and was separated from the Area B 4 x 6 m excavation by a 1 m baulk. Each area contained a single well defined layer of midden (Layer 2), under which lay another two cultural layers containing evidence of structures and minor amounts of shell (Layers 3 and 4) (Figure 6.7). Stratigraphic analysis showed the Layer 2 middens were contiguous, though Area B had a greater concentration of midden deposit. Beneath Layer 2 in areas A and B, a broken lens of clay/soil mix (Layer 3) was identified in which drain/ditch features extending from the south-west corner of Area B into the north-east corner of Area A were found. A hangi was also identified in this layer in Area A, and two small fire places (possibly hearths) in Area B. In Area A Layer 4 (natural), two pits were uncovered, each associated with drains and one with a series of postholes. A large homestead built over the site in European times was represented by three large postholes found in the bottom layer of Area A (Grey 1994).

Grey (1994:5) interpreted the prehistoric occupation sequence on the terrace as beginning with the construction of pits in Area A (Layer 4). This was followed by filling of the pits in Area A at a later time, the building of hangi in Areas A and B, and the digging of drain/ditches across A and B (Layer 3). Midden then built up across both areas (Layer 2) concentrated mainly in Area B. Finally, a homestead was built over the site in historic times (Grey 1994:5). The midden was dominated by the shellfish species *Paphies australis* and *Austrovenus stutchburyi*, with lesser amounts of *Paphies subtriangulatum*. *Pseudocaranx georgianus* was the dominant fish species exploited in Areas A and B, followed by *Arrispis trutta* and *Pagrus auratus*, with *Parika scaber* (Kokiri) being of minor importance (Kerr 1994:19).

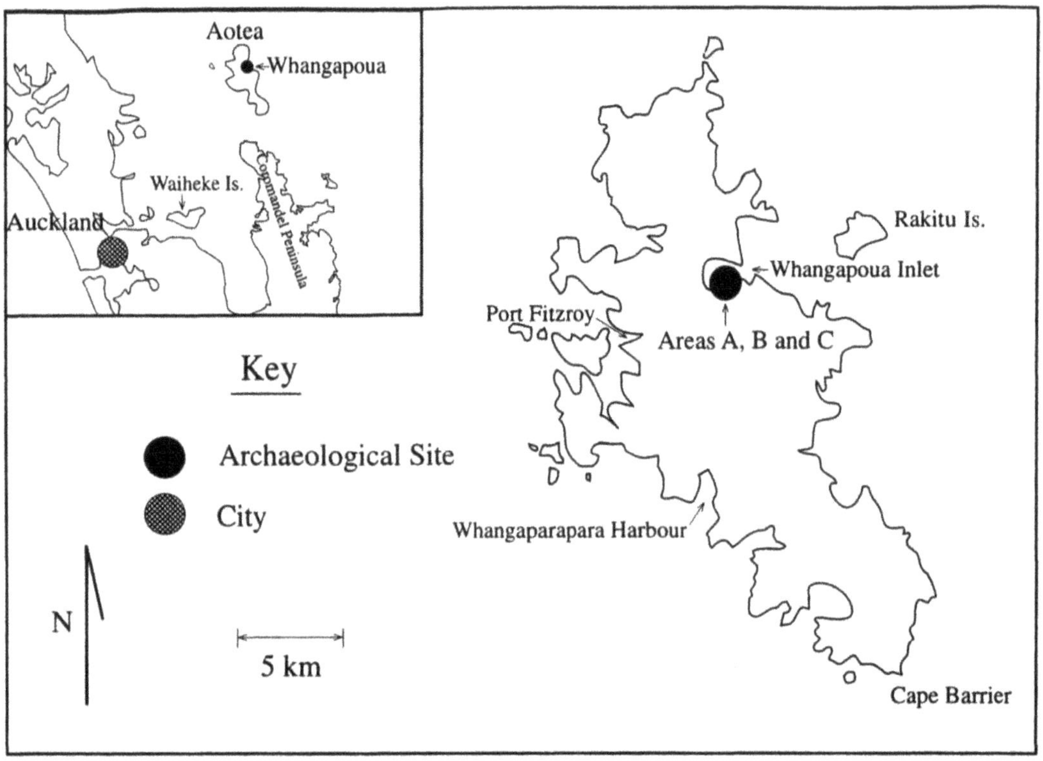


Figure 6.5. Location of Areas A, B (site S8/171) and C (site S8/16) on Aotea.

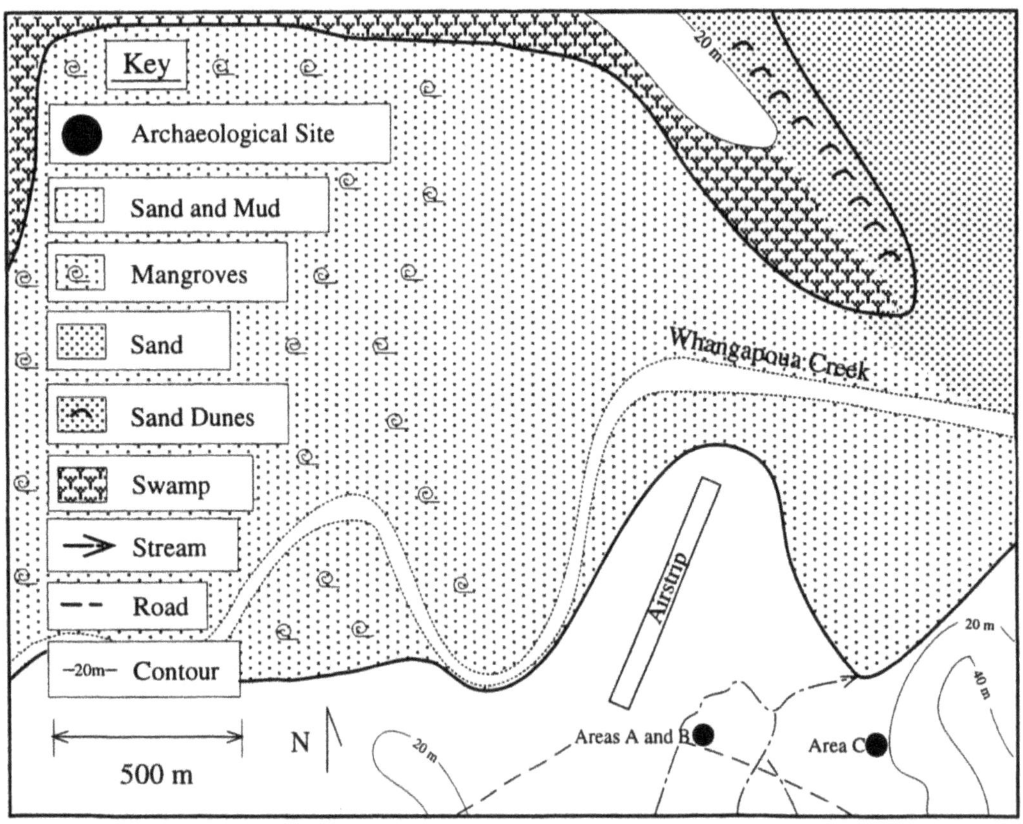


Figure 6.6. Location of Areas A, B (site S8/171) and C (site S8/16) by the Whangapoua estuary, Aotea.

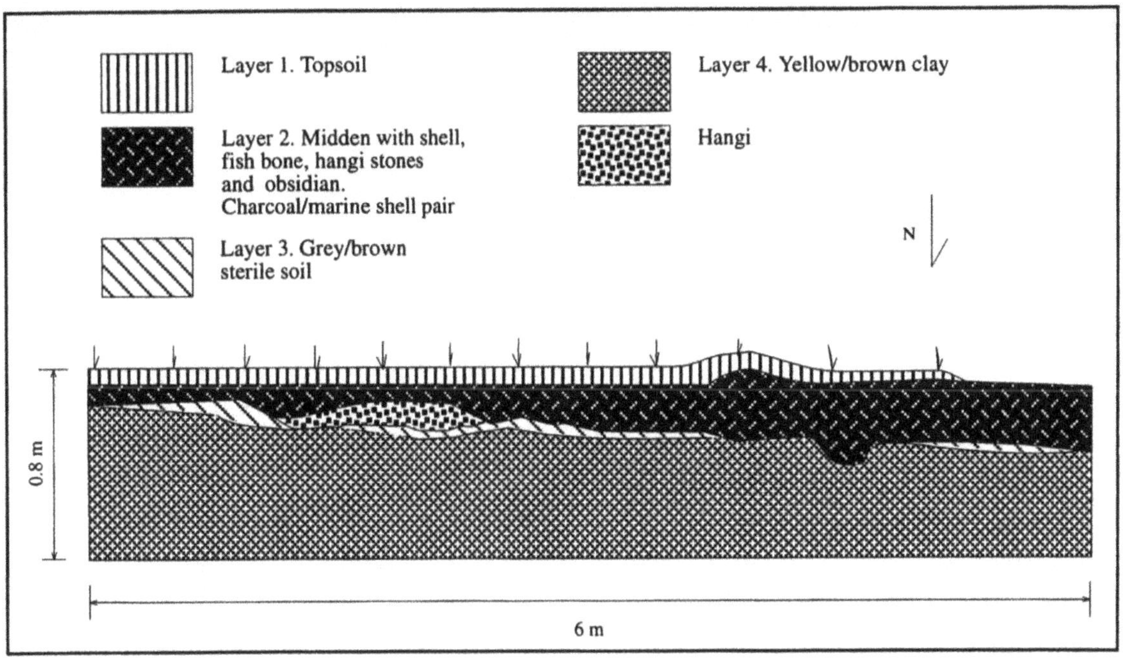


Figure 6.7. Stratigraphy of Area B, southern baulk (equivalent to stratigraphy of Area A northern baulk), site S8/171, Whangapoua, Aotea.

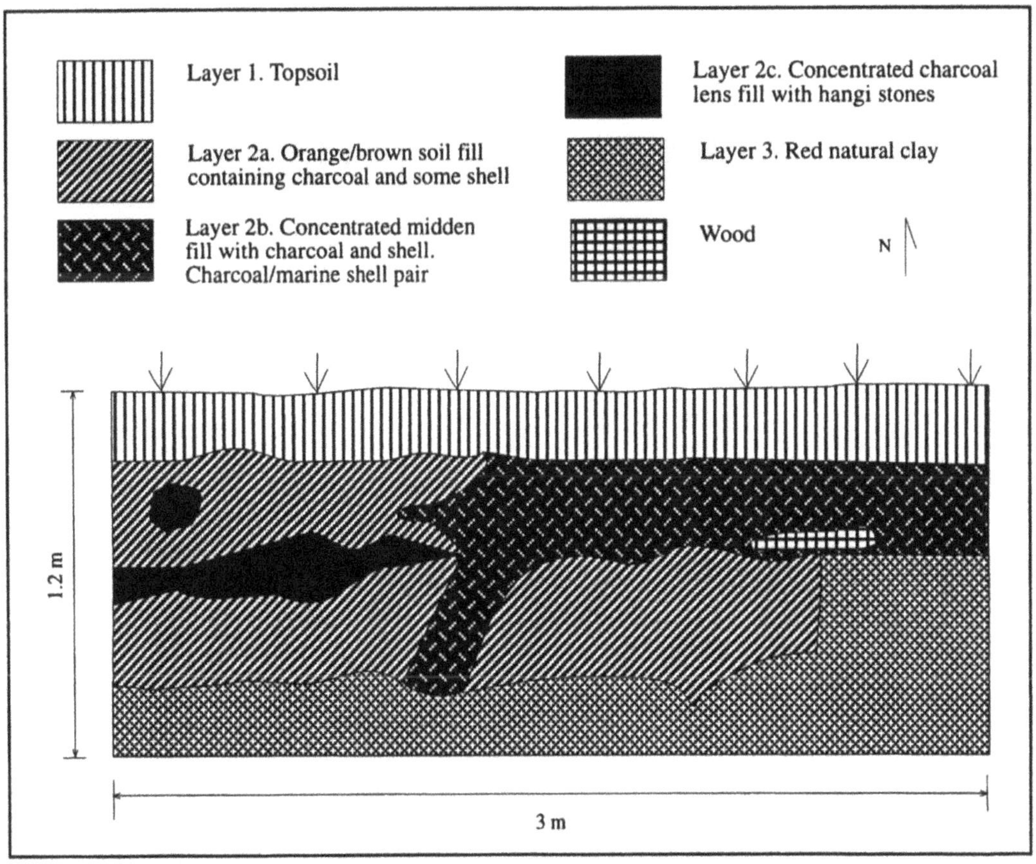


Figure 6.8. Stratigraphy of Area C, northern baulk, site S8/16, Whangapoua, Aotea.

Many hangi stones were also found in the midden, and, near the base of Layer 2 in Area A, three adzes and 350 obsidian flakes (Grey 1994:3; Elzenaar 1994:16).

Samples for dating were taken from Layer 2 of both areas to confirm time of occupation of the sites and to compare with obsidian hydration ages from the same provenance.

The Area C 3 x 6 m excavation yielded two cultural events: the construction and use of a pit (Layer 3), followed by filling of the pit using soil and midden to level the terrace for occupation (Layer 2) (Figure 6.8) (Grey 1994:7). The Layer 3 pit was dug into the natural soil substrate, and included a drainage system and associated postholes. The pit and terrace fill (Layer 2) contained shell, charcoal and hangi stones from cooking activities. Shellfish species from Layer 2 were dominated by *Paphies australis* and *Austrovenus stutchburyi*. *Paphies subtriangulatum* was a minor element of the midden. The main fish species exploited was *Pseudocaranx georgianus*, the midden also containing smaller amounts of *Pagrus auratus* and *Parika scaber* (Leather Jacket) (Kerr 1994:19). Obsidian accounted for the artifactual material recovered and probably came from both Mayor Island and Aotea sources. Samples for dating were taken from Layer 2 to determine when occupation of the terrace had occurred.

## Environmental Reconstruction

In contrast to the dense forest surrounding the Kokohuia site, Whangapoua represents a site immediate to scrub and coastal species. These include forest fringe and scrub species such as *Kunzea ericoides* and *Leptospermum scoparium* (dominates), and *Dodonaea viscosa, Coprosma, Olearia* and *Hebe* genus species (co-dominant). The common estuarine plant *Avicennia marina* var. *resinifera* was also identified. Larger kauri-podocarp-hardwood species such as *Agathis australis* and *Beilschmiedia tarairi* indicate a forest inland from the sites. Historic records attest to the cutting of kanuka and manuka for firewood for Auckland city, thereby illustrating the widespread distribution of these species on the coastal margins (Harlow 1994:13). Thousands of hectares of kauri were extensively logged on European arrival mainly from the west coast of the island (Harlow 1994:15). The charcoal identifications from the sites as well as the historic records, therefore, indicate the sampled sites were close to forest edge species which were extensive along the eastern coast of the island. This fringe of kanuka and manuka then gave way to kauri-podocarp-hardwood forest in the interior.

From an analysis of the marine faunal assemblages, the modern marine environment appears to be similar to that in prehistory. *Paphies australis* and *Austrovenus stutchburyi* occur in close proximity to the site today as in the past. *Paphies subtriangulatum* are still available on the beach to the north of the sites. The low occurrence of this species in the prehistoric record, however, may be due to restricted availability on the opposite side of the estuary. Fish species caught today are also present in the site inferring consistent availability. The harbour environment was probably similar to today, although infilling of part of the north west section of the estuary by sediments through past logging and recent activities, has possibly reduced the size of the estuary (Preston-Thomas 1994:12).

The geological strata underlying the Whangapoua sites is typical of the whole island. The sites sit upon an Upper Quaternary high level terrace of alluvium forming the Hawera Series, which extends up to 75m above sea level (Thompson 1960). According to Fleming and Harris (1978:549-550), the Hawera Series shows evidence of warm and cold climate episodes, where littoral marine deposits were laid down with the rising and falling of sea levels. Lying beneath this alluvium and dominating the geology of Aotea, are the Miocene Beeson's Island Volcanics, containing andesitic breccias, tuffs and agglomerates (Thompson 1960; Watters 1978:639). Though not directly in association with the archaeological sites under investigation, the Moehau Formation is the only major outcrop of Jurassic deposits on the island, bordering the Miocene Beeson's Island Volcanics, and forming the tip of Aotea from the northern edge of Whangapoua beach onwards. This bed is dominated by dark lithic-volcanic greywacke sandstone, interbedded with argillites, shales and conglomerates of andesite and granite, and is carbonate free (Thompson 1960; Skinner 1978:267). Streams flowing through these deposits and into the estuary, therefore, present no risk of a hard water effect.

| Laboratory No. | Provenance | Sample Type | Conventional Age BP | Calibrated Age AD 1σ (Stuiver and Reimer 1993) |
|---|---|---|---|---|
| NZ-7764 | Layer 4 | Charcoal | 535 ± 48 | 1334 - 1451 |
| NZ-7765 | Layer 4 | *Mytilus edulis aoteanus* | 957 ± 39 | 1326 - 1419[21] |

Table 6.1. Radiocarbon dates from Layer 4, N43/1, Ponui Island.

[21] Shell CRA calibrated using the modelled marine curve (Stuiver and Braziunas 1993) and New Zealand ΔR value of -25 ± 15 (Higham and Hogg 1995).

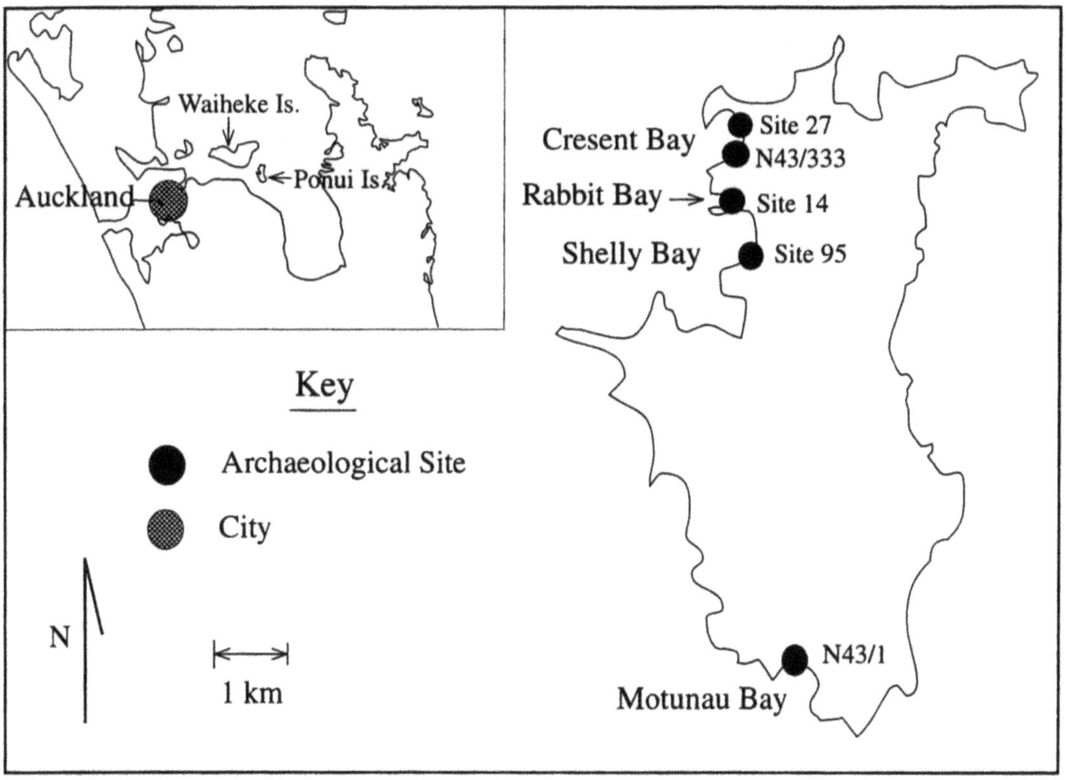


Figure 6.9. Locations of Ponui Island archaeological sites.

## PONUI ISLAND

### Archaeological Excavations and Analysis

Ponui Island lies 30 km southeast from Auckland City and is one of a number of islands in the Hauraki Gulf with extensive evidence of prehistoric activity (Figures 6.1 and 6.9). The island's prehistory spans from the Archaic through to Pakeha arrival (Irwin pers. comm.[22]). Five archaeological sites on the island (Figure 6.9) were sampled in November 1994 by Prof. G. Irwin, Dr. R. Wallace and Dr. T. Ladefoged of the Auckland University Anthropology Department, and the author. Wet sieving was undertaken to maximise the recovery of charcoal from the sites, allowing a large amount of charcoal to be retained. The stratigraphic sequence of all the Ponui archaeological sites comprised of midden deposits on a natural beach base, interspersed with layers of beach sand or slope wash.

Site N43/1 was located adjacent to an estuary mouth which opened out onto the long sandy beach of Motunau Bay (Figure 6.10). This site contained two cultural deposits, the lower of which (Layer 4) had been previously dated (Table 6.1) (Figure 6.11). These determinations yielded occupation at *ca.* 500 years BP. Layer 4 was a dense midden containing only shellfish and charcoal. This layer was excavated as three spits (A, B and C) as investigation revealed changes in shellfish frequency with depth. This possibly reflected a number of separate meals at the same location within a short period of time. *Paphies australis, Austrovenus stutchburyi, Paphies subtriangulatum, Crassostrea glomerata* and *Perna canaliculus* constituted the shellfish species present in Spit C, and all species were sampled for dating.

Site 27 was a beach front midden in Crescent Bay and contained two cultural deposits, both distinctive in character, but with no intervening deposit (Figure 6.12 and 6.13). The lower of the deposits was sampled (Layer 4). The Layer 4 midden had a thick layer of charcoal at its base, possibly resulting from a brief burn of surrounding scrub on or near the site, contemporary with deposition of the midden (Irwin pers. comm.). The shellfish species present in the midden were restricted to *Paphies australis* and *Austrovenus stutchburyi*. No fish bone was observed.

N43/333 was also located in Crescent Bay next to a stream mouth 20 m south of Site 27. This site had two cultural layers separated by a sterile layer of clay/beach sand mix with tailings of eroded shell (Figure 6.14). The lower deposit (Layer 4) contained not only midden but Archaic artefacts including a small Duff 1A adze, a small stone chisel, bird bone awl and 4 flakes of obsidian: three from Mayor Island (green) and the other perhaps from a Coromandel source (white). *Paphies australis* and *Austrovenus stutchburyi* were the most common shellfish exploited, but no fish bone was identified. The upper cultural deposit (Layer 2) lacked artefacts and contained *Paphies australis, Austrovenus stutchburyi,* and *Crassostrea glomerata*, but no fish bone. Both Layers 2 and 4 were sampled for dating.

[22] Prof. G. Irwin, Department of Anthropology, University of Auckland.

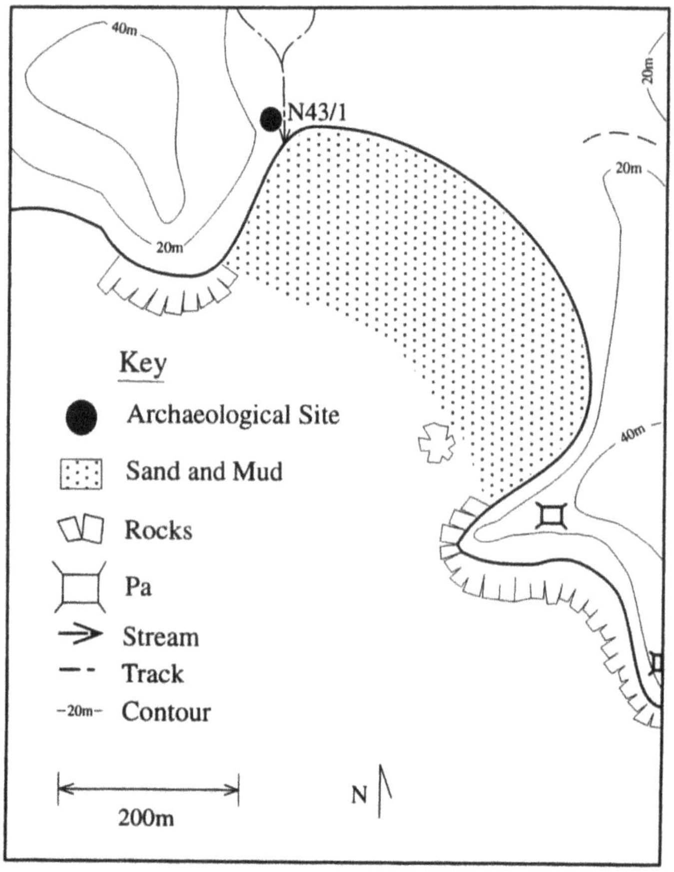


Figure 6.10. Location of site N43/1 in Motunau Bay, Ponui Island.

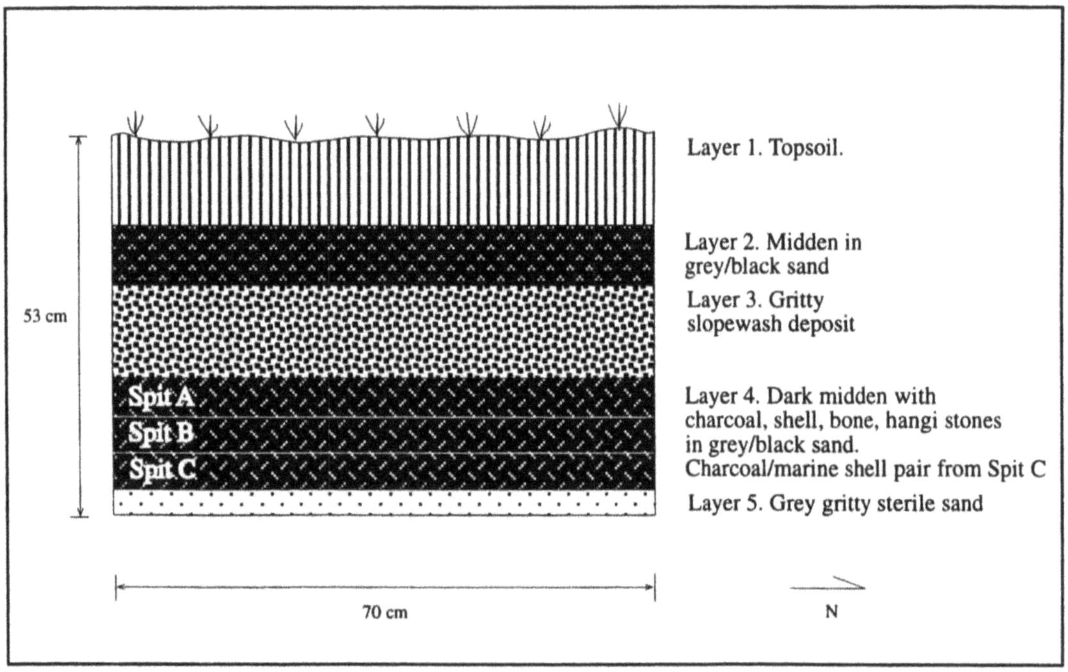


Figure 6.11. Stratigraphy of site N43/1, Motunau Bay, Ponui Island.

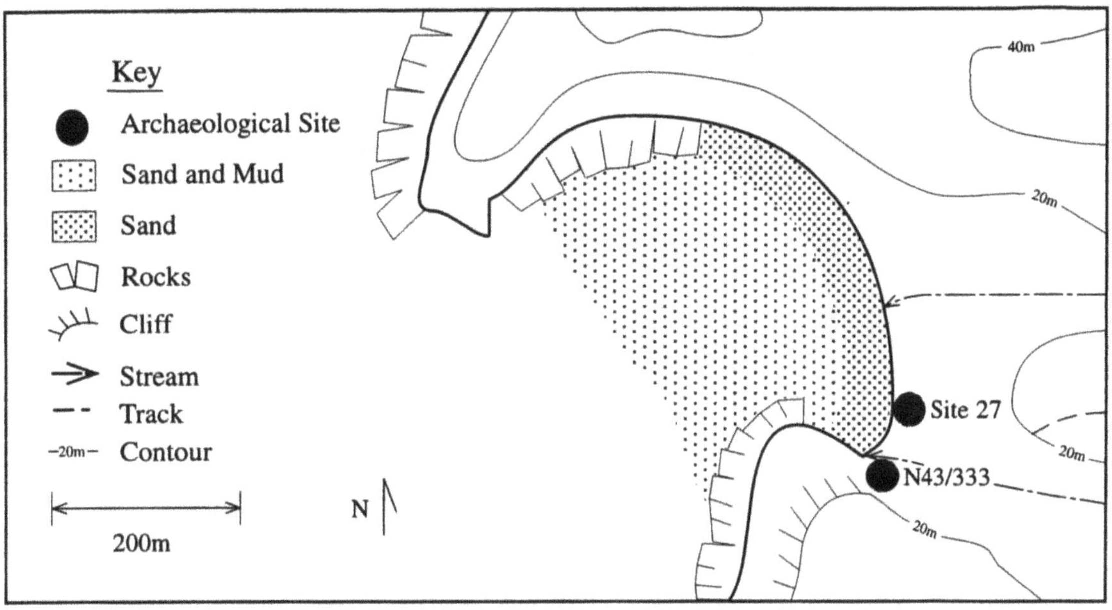


Figure 6.12. Locations of sites 27 and N43/333, Crescent Bay, Ponui Island.

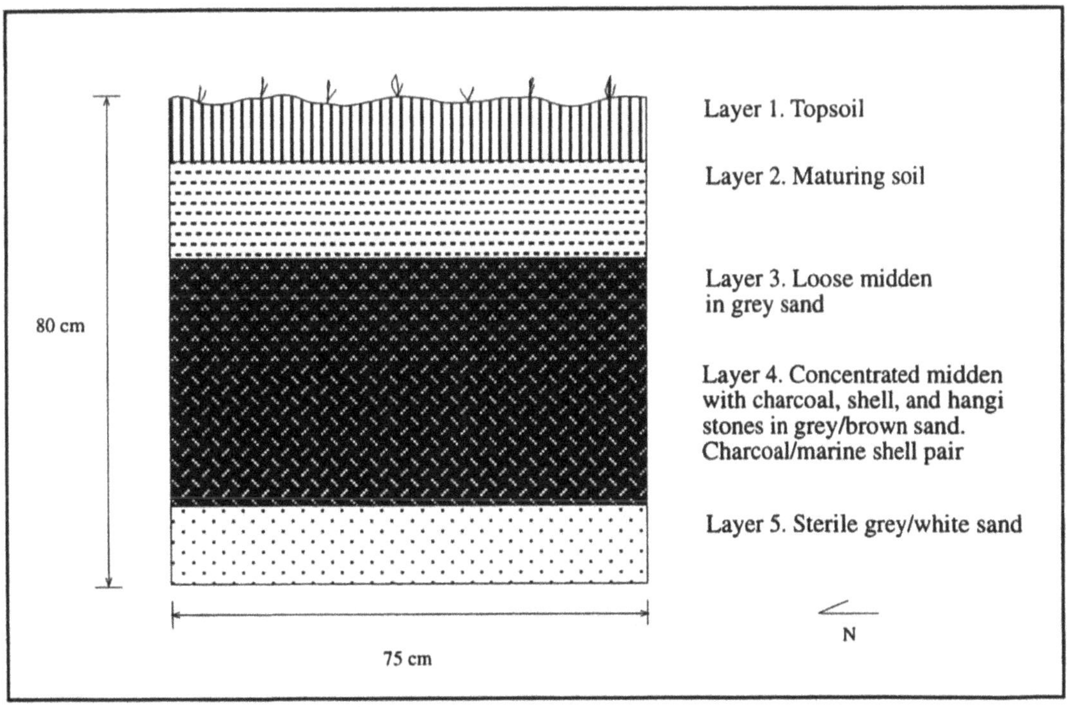


Figure 6.13. Stratigraphy of site 27, Crescent Bay, Ponui Island.

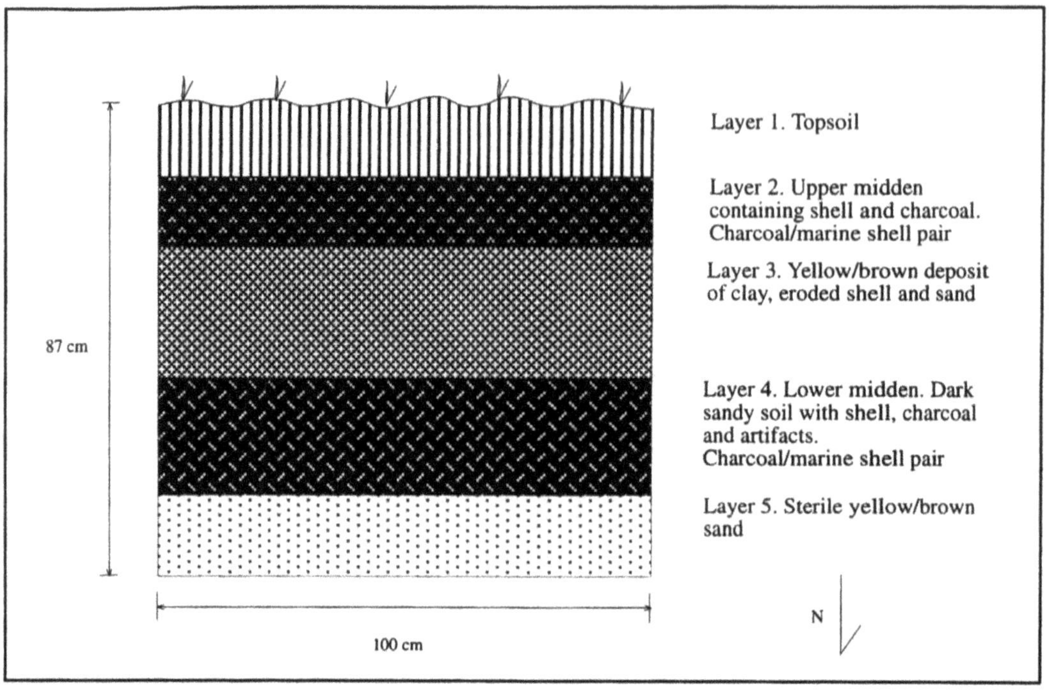


Figure 6.14. Stratigraphy of site N43/333, Crescent Bay, Ponui Island.

Site 14 was situated on the sandy beach of Rabbit Bay (Figure 6.15). This site contained three cultural layers, with Layer 4 being sampled for dating (Figure 6.16). Layer 4 was a dense clean midden within a yellow beach sand and was profuse in *Pagrus auratus* bone. *Paphies australis, Austrovenus stutchburyi, Crassostrea glomerata* and *Perna canalicula* were the shellfish species identified.

Site 95 was located next to a small stream flowing into Shell Bay (Figure 6.17). Two cultural deposits (Layers 2 and 4) were identified, each containing midden consisting of charcoal and *Austrovenus stutchburyi* (Figure 6.18). Layers 2 and 4 were separated by a natural beach deposit of grey sand and weathered shell fragments (Layer 3). Excavation of Layer 2 revealed a posthole which extended down into Layer 3 implying the building of a structure on the beach front in prehistory. The Layer 4 midden was identical in appearance to Layer 2, but contained no hangi stones.

Archaeological deposits containing only one shellfish species are an unusual feature of prehistoric middens on Ponui as seen in the above excavations. Because of this, and the close similarities in appearance of the deposits, Wallace[23] and Irwin (pers. comm.) suggested that Layers 2 and 4 could represent a mixture of natural beach shell and culturally derived midden. It was therefore decided to sample each layer to investigate this possibility.

### Environmental Reconstruction

In prehistoric times, the forest edge and coastal species of *Kunzea ericoides, Leptospermum scoparium, Pseudopanax arboreus* (Puahou), *Dysoxylum spectabile* (Kohekohe), *Lophomyrtus bullata* (Ramarama) and *Metrosideros excelsa* (Pohutukawa) were adjacent to all sites on Ponui. Species from the genus *Coprosma, Hebe,* and *Olearia* were also present. Closely bounding the sites was *Agathis australis* forest with associated medium sized understorey trees of *Beilschmiedia tarairi* and *Metrosideros robusta* (Vine Rata). The Ponui sites, therefore, appear to be directly associated with a dense forest similar to that found at Kokohuia.

Large *Metrosideros excelsa* and forest edge species are still present today at all the sites excavated. The great kauri forests seen on European arrival, however, are long gone due to logging (Romanes 1993).

All the bays on Ponui have the same shellfish species available today as in prehistory. Snapper was the only fish species identified from the excavations (at Site 14), and this species is still fished today. Due to Ponui Island's location within a major seaway, and the open nature of its bays, tidal flushing is efficient. This was probably the case in prehistory, though an increase in sedimentation of the bays from early Pakeha forestry may have reduced their size (as at Aotea).

Ponui is a small hilly island containing many bays along its coast. The sites sampled lie within recent beach deposits, which rest upon Holocene Hawera alluvia. Under these alluvia are Upper Jurassic sandstone and mudstones of the Chamberlin and Mangatangi Formations which are found throughout the island, except at Rabbit Bay (Schofield 1979:4-5). The Mangatangi and Chamberlin Formations are described by Schofield (1978:265;1979:4) as having an absence of macrofossils and microfaunas. These strata extend out into the Waiheke Channel and Tamaki Strait forming the shallow bays on the western side of the island.

[23] Dr. R. Wallace, Department of Anthropology, University of Auckland.

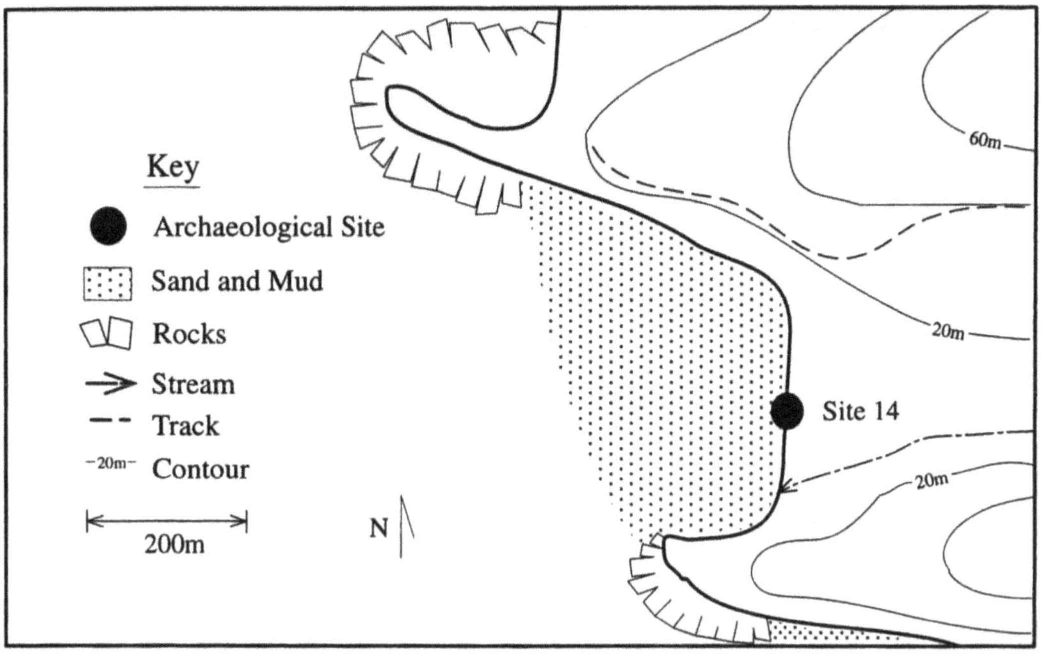


Figure 6.15 Location of site 14, Rabbit Bay, Ponui Island.

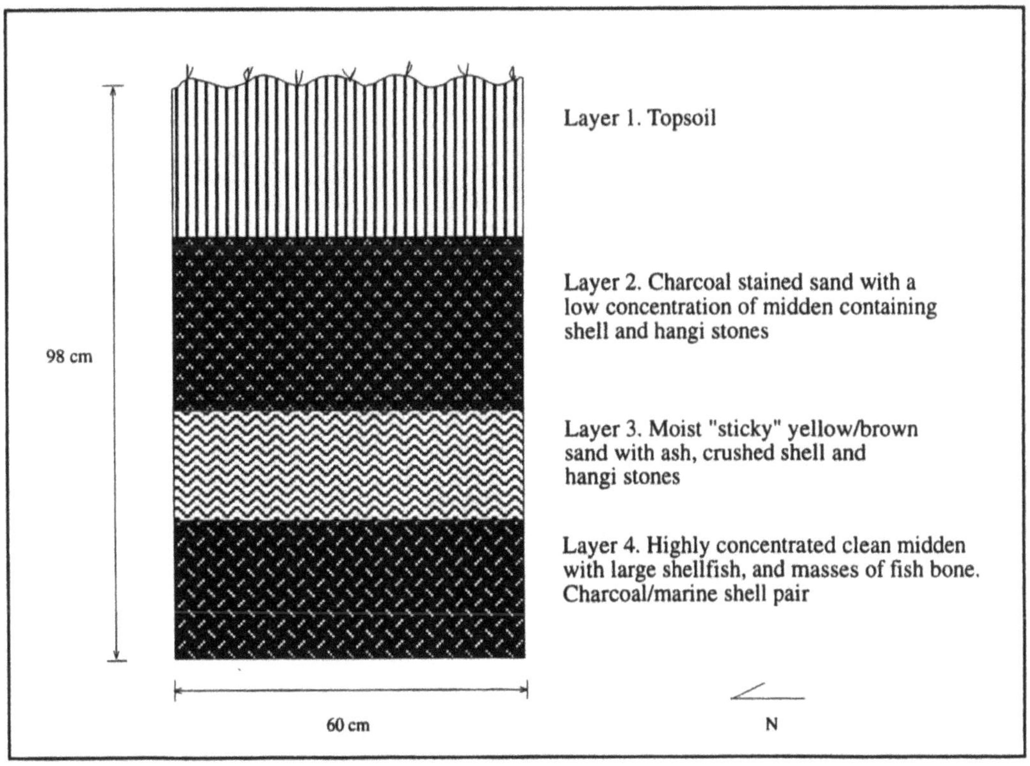


Figure 6.16. Stratigraphy of site 14, Rabbit Bay, Ponui Island.

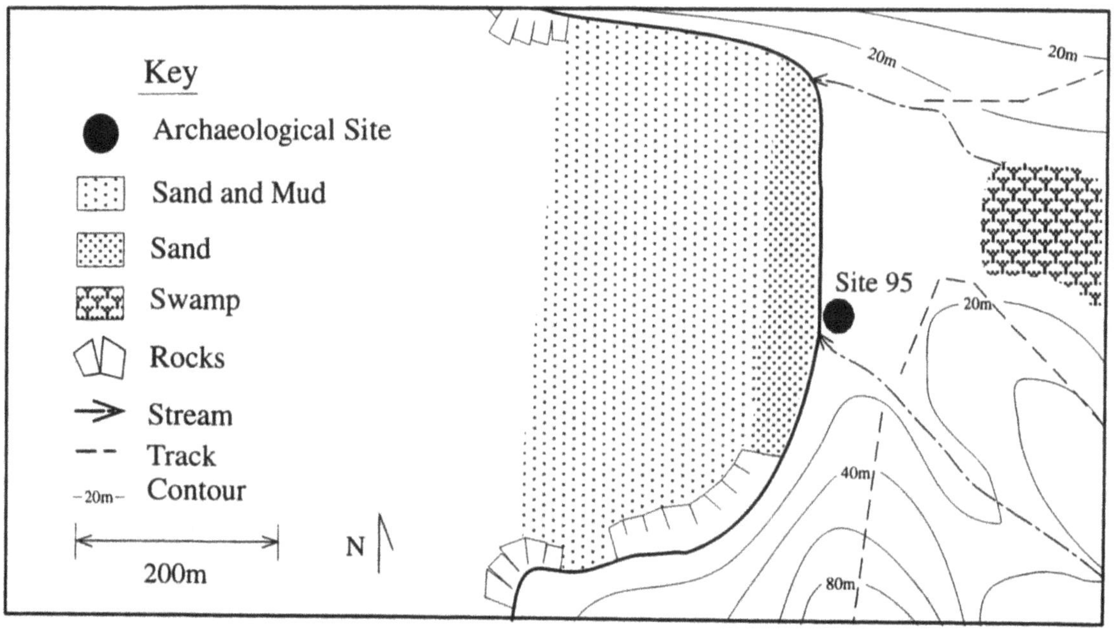


Figure 6.17. Location of site 95, Shell Bay, Ponui Island.

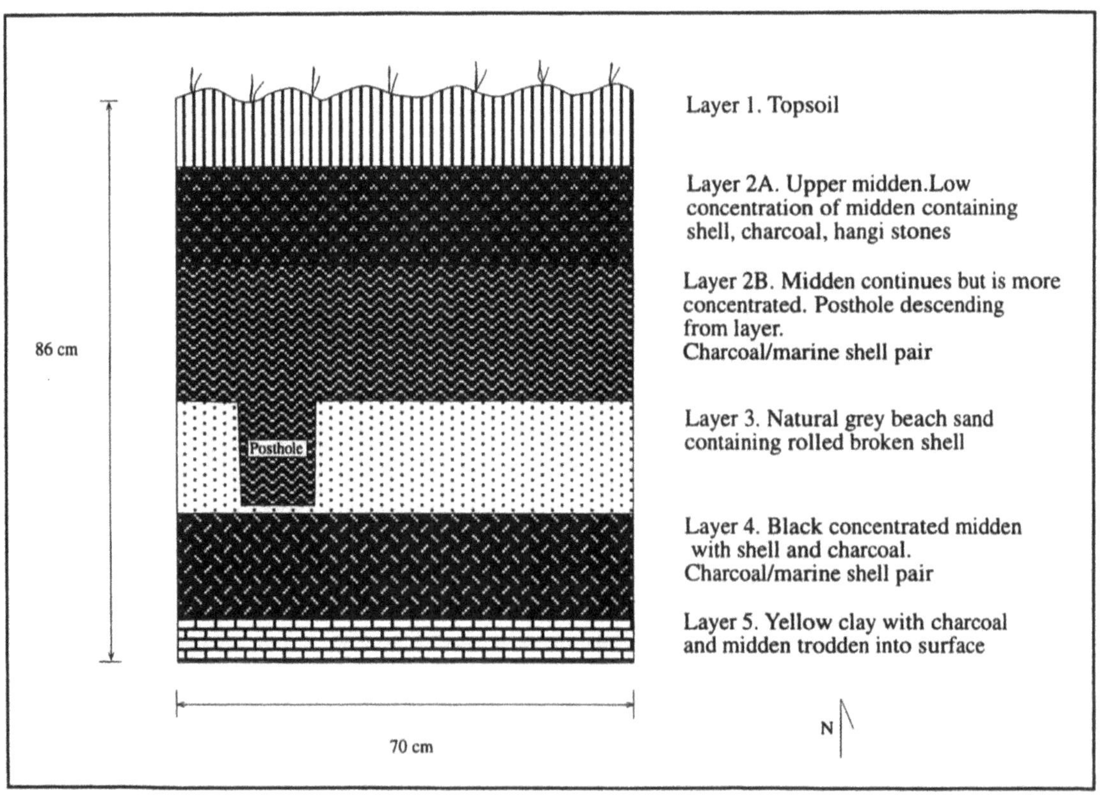


Figure 6.18. Stratigraphy of site 95, Shell Bay, Ponui Island.

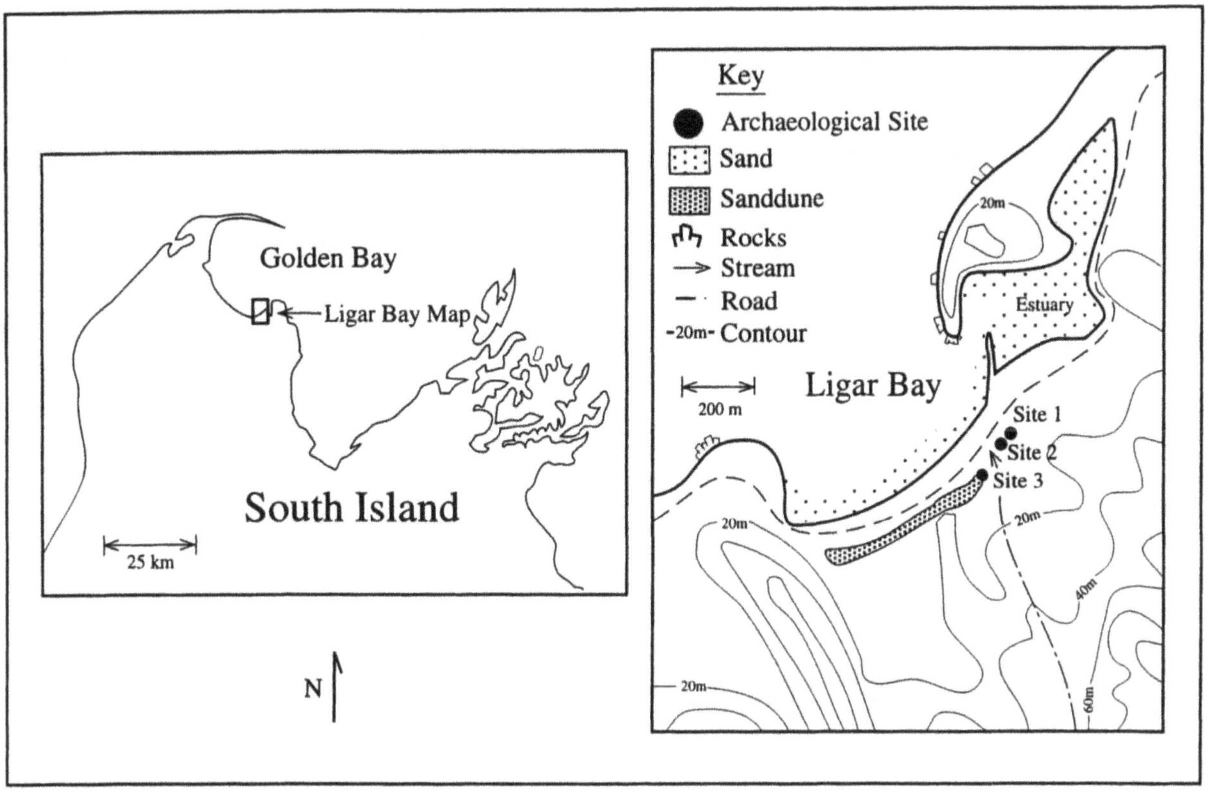


Figure 6.19. Location of Ligar Bay archaeological sites, Golden Bay.

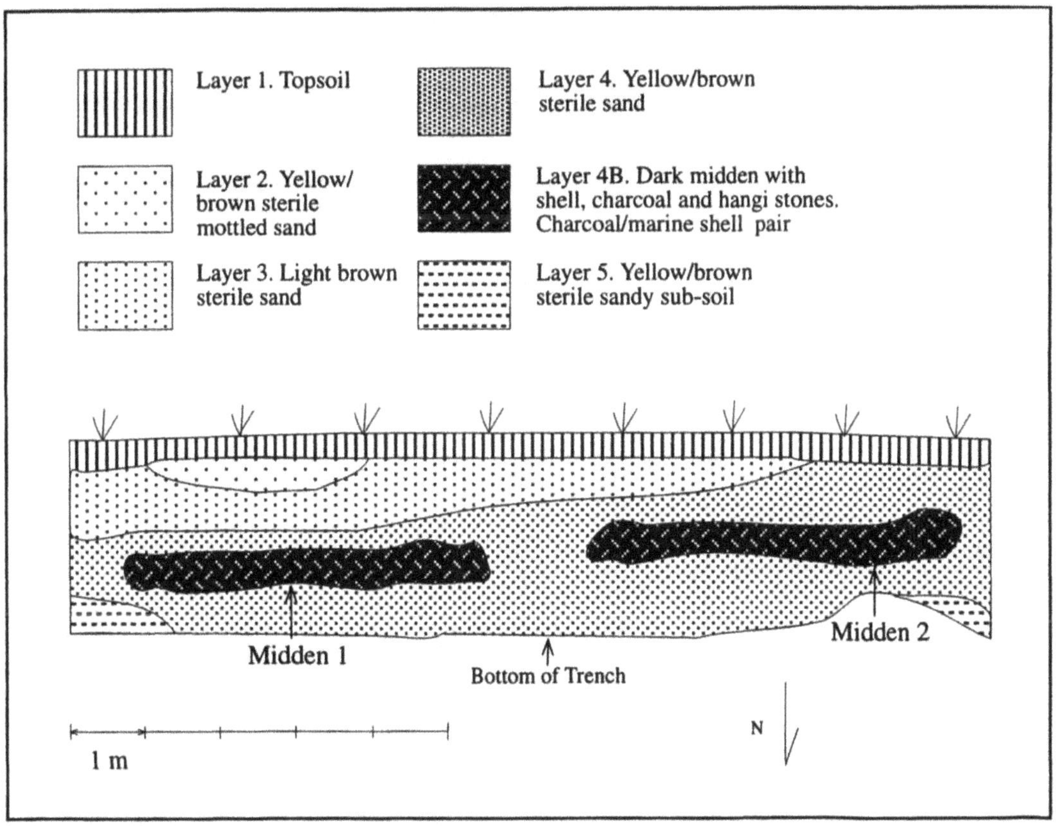


Figure 6.20. Stratigraphy of site N25/95, Middens 1 & 2, Ligar Bay.

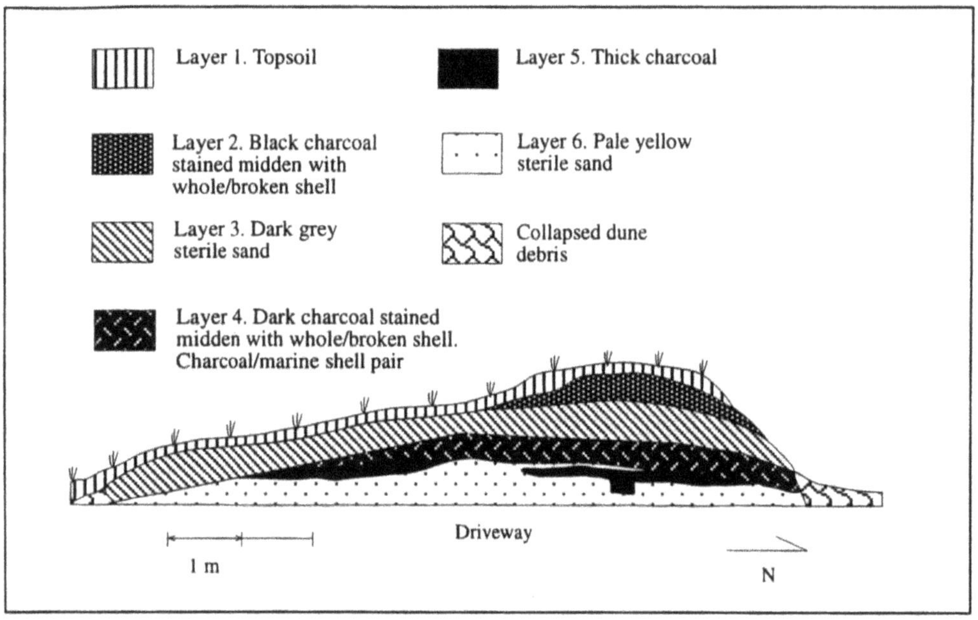


Figure 6.21. Stratigraphy of site N25/26, Ligar Bay.

## LIGAR BAY

### Archaeological Excavations and Analysis

Ligar Bay is situated within Golden Bay at the top of the South Island (Figure 6.1). Two sites lying within 20 m of the present high tide mark were exposed during sub-division activities on farm land (Figure 6.19). This find required the sites to be surveyed and assessed by W. Gumbley of the New Zealand Historic Places Trust in August 1994, giving the opportunity to sample three distinctive archaeological deposits in close proximity to each other.

Each site was located on a beach front in prehistory, one on the beach flat, and one on a sand dune. Samples were taken from two middens present in the beach flat site (N25/95, Middens 1 and 2), now exposed 0.5 m below the present ground surface, and from an upper occupation level from a surviving portion of the dune (N25/26, Layer 4).

Midden 1 of site N25/95 measured approximately 4.5 x 2 x 0.5 m, and contained shell, fish bone, hangi stones and charcoal (Figure 6.20). This deposit lay as a lens within a yellow/brown granitic sand. *Paphies australis, Austrovenus stutchburyi* and *Paphies subtriangulatum* were the dominant shellfish species present, followed in lesser amounts by *Lunella smaragda, Mytilus edulis aoteanus* and *Amphibola crenata.* Fish species identified were *Pseudophycis bachus* (Red Cod), *Thyrsites atun* (Barracouta) and *Genypterus blacodes* (Ling). The dimensions and isolation of this midden suggested a single event.

Midden 2 was located within 2 m of midden 1 and was similar in content, containing shell, fishbone, hangi stones and charcoal (Figure 6.20). This deposit was also sealed by a yellow/brown granitic sand. Common shellfish species present were *Paphies australis, Austrovenus stutchburyi, Paphies subtriangulatum, Lunella smaragda, Mytilus edulis aoteanus* and *Amphibola crenata.* Identified fish were *Pseudophycis bachus, Thyrsites atun* and *Genypterus blacodes.* An adze flake made possibly from D'Urville Island argillite was also found. The extent of the deposit (approx. 5 x 2 x 0.4 m) implied another single prehistoric event.

The events at Middens 1 and 2 were interpreted as occurring at the same time due to their close proximity, similar vertical orientation, and their isolation within a sterile sand layer. This interpretation was also supported by close similarities in shellfish variety and size, and common fish species.

Layer 4 of site N25/26 was a compact, solidified midden, containing whole and crushed shell, fish bone, hangi stones and charcoal (Figure 6.21). Shellfish species consisted of *Paphies australis, Austrovenus stutchburyi, Paphies subtriangulatum, Lunella smaragda* and *Cominella adspersa.* Fish species present were *Pseudophycis bachus, Thyrsites atun, Genypterus blacodes* and *Pagrus auratus.* The layer sampled (Layer 4) was one of four distinctive cultural events.

### Environmental Reconstruction

The charcoal identifications indicate that a forested environment surrounded all the sites in Ligar Bay. *Dodonaea viscosa, Myrsine australis, Kunzea ericoides, Coprosma* sp., *Olearia,* sp., *Pittosporum* sp. and *Hebe* sp. were the coastal and forest edge species identified. *Nothofagus* (Beech) (dominant), *Podocarpus totara, Prumnopitys taxifolia* (Matai), *Weinmannia racemosa* (Kamahi), *Metrosideros umbellata* (Southern Rata) and the smaller under canopy species of *Melicytus ramiflorus* (Mahoe) and *Macropiper excelsum* (Kawakawa) forested the hillsides. Crops of

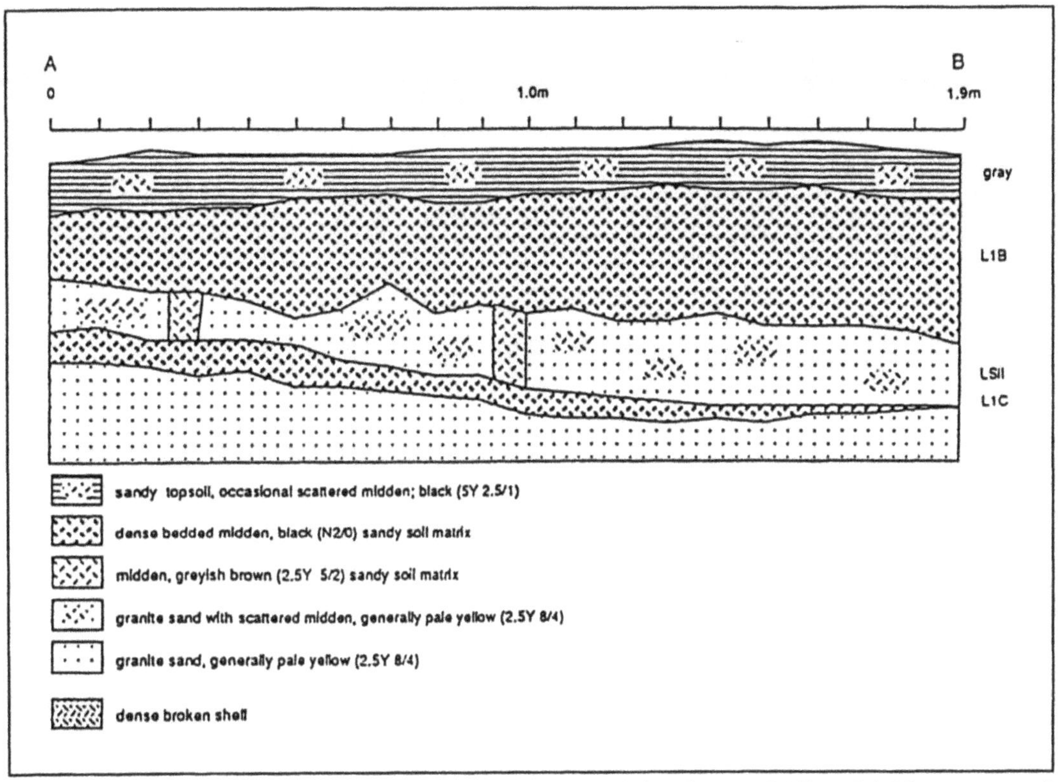


Figure 6.23. Stratigraphy of Sawpit Point (N26/214), Awaroa Inlet. Layers L1B and L1C were sampled for dating.

| Laboratory No. | Provenance | Sample Type | Conventional Age BP | Calibrated Age AD 1σ (Stuiver and Reimer 1993) |
|---|---|---|---|---|
| NZ7701i | L1C | Charcoal | 350 ± 110 | 1450 - 1800 |
| Wk-3099 | L1C | *Paphies australis* | 800 ± 50 | 1451 - 1536[24] |

Table 6.2. Radiocarbon dates from Layer L1C, Sawpit Point (N26/214), Awaroa Inlet.

*Pteridium aquilinum* var. *esculentum* (Bracken fern) stood in forest clearings. This landscape is similar to that surrounding Kokohuia and Ponui in prehistory, and is typical of South Island coastal Beech Forest. This terrestrial environment was still present on European arrival (Barber 1994:952-954).

The shellfish species from the sites indicate a similar marine environment today as in the past, with *Paphies australis, Austrovenus stutchburyi* and *Paphies subtriangulatum* all being available. *Pseudophycis bachus, Thyrsites atun, Genypterus blacodes* and *Pagrus auratus* are also caught today.

The Ligar Bay sites sit within Holocene sands of granitic origin which lie on Bainham Formation Quaternary gravel deposits (Grindley 1971:16-17). Beneath these gravels is the Separation Point Batholith of granodiorite and other associated minerals of Lower Cretaceous-Jurassic origin. It is this feature that dominates the region from Ligar Bay eastwards as far as Motuaka (Grindley 1971:10-12; Grindley 1978: Fig. 2.13, 95-96; Barber 1994). Bordering the very western edge of Ligar Bay is a small outcrop of Takaka Limestone, and on the western and eastern margins traces of the Motupipi Coal Measures (Grindly 1971:15). These deposits sit as a lens upon the Separation Point Batholith, and represent little risk as sources of hardwater for the bay.

## SAWPIT POINT (N26/214)

### Archaeological Excavation and Analysis

The Sawpit Point site (N26/214) was located in Awaroa Inlet along the Granite Coast east of Ligar Bay, and was excavated by Barber from 1988-89 (Figure 6.1). This site was part of a dune complex containing midden and pits. The midden deposit extended over several hundred square meters (Barber 1994:210). The excavation involved investigation of a pit and surrounding cultural deposits to ascertain whether the pits had been used for food storage, and determine the relationship between the pits and the activities reflected in the *in situ* midden deposits (Barber 1994:211).

[24] Shell CRA calibrated using the modelled marine curve (Stuiver and Braziunas 1993) and New Zealand ΔR value of -25 ± 15 (Higham and Hogg 1995).

The excavation uncovered three distinctive cultural deposits, each separated by sterile granite sand and intermittent midden concentrations (Barber 1994:211) (Figure 6.23). These stratigraphic units were disturbed only by the construction and infilling of a pit. *Paphies australis* dominated the shellfish species present in all layers. *Austrovenus stutchburyi*, *Amphibola crenata*, *Lunella smaragda* and *Cominella* sp. occurred in lesser amounts. *Pseudophycis bachus* and *Thyrsites atun* made up almost 90% of fish species identified. Mammal and bird remains consisted of *Canis familiaris*, *Rattus exulans*, *Delphinus delphis* (Seal), *Puffinus gavia* (Shearwater), *Eudyptula minor* (Little Blue Penguin) and *Callaeas cinerea* (Kokako). Artefacts recovered included four bone bird spear points, three bone fish hooks, and slender bone shafts, all made from dog bone (Barber 1994:611). Dog coprolites were also identified in the midden.

Two initial radiocarbon dates from Layer L1C were run by the Waikato and IGNS laboratories to date first occupation on the dune (Table 6.2). The large standard error associated with the IGNS charcoal age was due to laboratory error (Barber 1994:228; Barber pers. comm.[25]). The shell date, though, can be considered to date the initial occupation event on the dune. Because the site was of a similar type to those investigated at Ligar Bay, but set well apart geographically, there was an opportunity to investigate the dating of marine shell on another section of this coastline from an already dated context. Layer L1C was resampled, and additional material was collected from L1B for dating.

Barber has noted that sub-fossil *Austrovenus stutchburyi* may be present in layer L1C (Barber 1994:220; Barber pers. comm.). This shell may have intruded into the midden during construction of the pit, and so radiocarbon ages on *Austrovenus stutchburyi* from L1C may be very old.

### Environmental Reconstruction

The forest cover present during prehistoric occupation at Sawpit Point was similar to the Ligar Bay sites. Barber (1994:952) identified *Nothofagus* as the dominant forest species immediately surrounding the site with *Metrosideros robusta, Leptospermum scoparium, Kunzea ericoides*, *Dodonaea viscosa* and *Melicytus ramiflorus* being the common coastal species (Barber 1995:953-954). He noted that the forest cover today is probably similar to that present before European arrival, other than some areas containing regenerating scrub (Barber 1994:954).

Like the Ligar Bay sites, Sawpit Point has a similar marine environment today as it did in the past, with all faunal species identified still being present. The geology of the Sawpit Point site is identical to that of the Ligar Bay sites, excluding the minor presence of the Westhaven Group Takaka Limestone and Motupipi Coal Measures. This site sits within a granitic Holocene sand dune upon the Separation Point Batholith and so represents a calcareous free deposit (Grindley 1971:10-12, 15-17; Barber 1994:187, 211).

## SUMMARY

The archaeological sites sampled above provide a cross-section of environments from which various shellfish species were resourced by Maori in prehistory. Dating these sites will enable the investigation of environmental influences upon shell dates, and ascertain whether there are species specific factors inherent in accurate marine shell dating. We may also observe how charcoal ages reflect the age of archaeological events when compared to marine shell dates, and what factors may increase the inbuilt age of identified charcoal samples.

[25] Dr. I. Barber, Historic Places Trust, Wellington.

# CHAPTER 7

# ANALYSIS OF RADIOCARBON AGES FROM EXCAVATED ARCHAEOLOGICAL SITES

## INTRODUCTION

In this chapter, the analyses of the paired charcoal and marine shell radiocarbon ages from the archaeological sites excavated and sampled for this dissertation are presented. Each marine shell and charcoal CRA from the archaeological sites have been calibrated separately, so the effect of the various calibration curves on the calibration of individual radiocarbon dates may be assessed.

Marine shell and charcoal radiocarbon dates are seen as being in agreement in dating an archaeological deposit, when age ranges from calibrated shell dates overlap with age ranges from calibrated charcoal dates at 1σ (see Kirch and Hunt 1988; Kirch, Hunt and Tyler 1989; Dye 1990; Anderson 1991; Allen 1994; Anderson *et al.* 1996). Archaeological sites where calibrated marine shell dates are in agreement with calibrated charcoal dates, when using either the reservoir correction calibration method, or the modelled marine curve calibration technique, are considered first in this analysis.

Archaeological sites where some of the calibrated charcoal and marine shell radiocarbon ages from the same provenance do not agree, irrespective of which marine shell calibration method is used, are then examined. For these sites, reasons are suggested for the age differences between the sample types.

For each of the archaeological sites discussed, evaluations are made on:

- similarities in radiocarbon ages between shellfish species;
- variability in radiocarbon determinations of charcoal samples from the same provenance;
- the effect of using either of the two marine shell CRA calibration methods for accurately depicting the age of an archaeological deposit.

## ARCHAEOLOGICAL SITES WITH CHARCOAL/MARINE SHELL CHRONOLOGICAL AGREEMENT

### Kokohuia (O06/317)

Charcoal and marine shell radiocarbon ages from Layers 2 to 4 from Kokohuia are presented below (Tables 7.1 - 7.4; Figures 7.1 - 7.4).

### Layer 2

| Lab No. | Sample Type | CRA | Cal Age AD 1σ (Atmospheric Curve) | Cal Age AD 1σ (Modelled Marine Curve) |
|---|---|---|---|---|
| Wk-3695 | Charcoal | 330 ± 45 | 1525 - 1562<br>1629 - 1658 | |
| Wk-4484 | Charcoal | 380 ± 40 | 1487 - 1642 | |
| Wk-4485 | Charcoal | 340 ± 40 | 1523 - 1578<br>1626 - 1653 | |
| Wk-4486 | Charcoal | 410 ± 40 | 1450 - 1528<br>1558 - 1631 | |
| Wk-4487 | Charcoal | 380 ± 40 | 1487 - 1642 | |
| Wk 3696 | *Paphies australis* | 710 ± 40 | 1493 - 1602<br>1616 - 1646 | 1523 - 1646 |
| Wk-3697 | *Austrovenus stutchburyi* | 770 ± 40 | 1445 - 1522<br>1583 - 1625 | 1479 - 1545 |
| Wk-3698 | *Venerupis largillierti* | 680 ± 40 | 1525 - 1562<br>1629 - 1654 | 1549 - 1663 |
| Wk-3699 | *Lunella smaragda* | 760 ± 40 | 1448 - 1526<br>1561 - 1629 | 1486 - 1560 |
| Wk-3701 | *Perna canaliculus* | 800 ± 40 | 1439 - 1492 | 1462 - 1520 |

Table 7.1 Conventional radiocarbon dates and calibrated ages on charcoal and marine shell from Kokohuia, Layer 2.

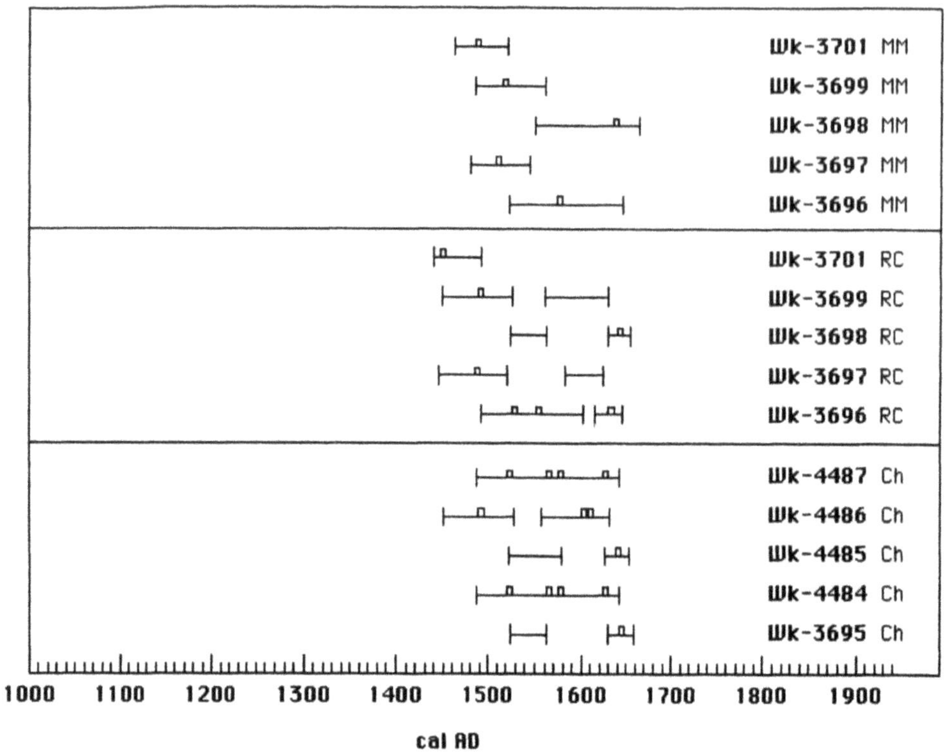


Figure 7.1. Calibrated radiocarbon ages from Kokohuia (O06/317), Layer 2. (MM = shell calibrated by Modelled Marine Curve; RC = shell calibrated using Reservoir Correction method; Ch = calibrated charcoal age).

## Layer 3

| Lab No. | Sample Type | CRA | Cal Age AD 1σ (Atmospheric Curve) | Cal Age AD 1σ (Modelled Marine Curve) |
|---|---|---|---|---|
| Wk-3703 | Charcoal | 410 ± 40 | 1450 - 1528<br>1558 - 1631 | |
| Wk-4540 | Charcoal | 360 ± 40 | 1495 - 1513<br>1516 - 1599<br>1618 - 1648 | |
| Wk-4541 | Charcoal | 350 ± 40 | 1519 - 1594<br>1622 - 1650 | |
| Wk-3704 | *Paphies australis* | 710 ± 40 | 1493 - 1602<br>1616 - 1646 | 1523 - 1646 |
| Wk-3705 | *Austrovenus stutchburyi* | 750 ± 40 | 1451 - 1530<br>1554 - 1633 | 1492 - 1587 |
| Wk-3706 | *Venerupis largillierti* | 720 ± 40 | 1489 - 1643 | 1516 - 1639 |
| Wk-3707 | *Lunella smaragda* | 750 ± 40 | 1451 - 1530<br>1554 - 1633 | 1492 - 1587 |
| Wk-3709 | *Perna canaliculus* | 740 ± 40 | 1455 - 1455<br>1478 - 1534<br>1539 - 1637 | 1500 - 1619 |

Table 7.2. Conventional radiocarbon dates and calibrated ages on charcoal and marine shell from Layer 3, Kokohuia.

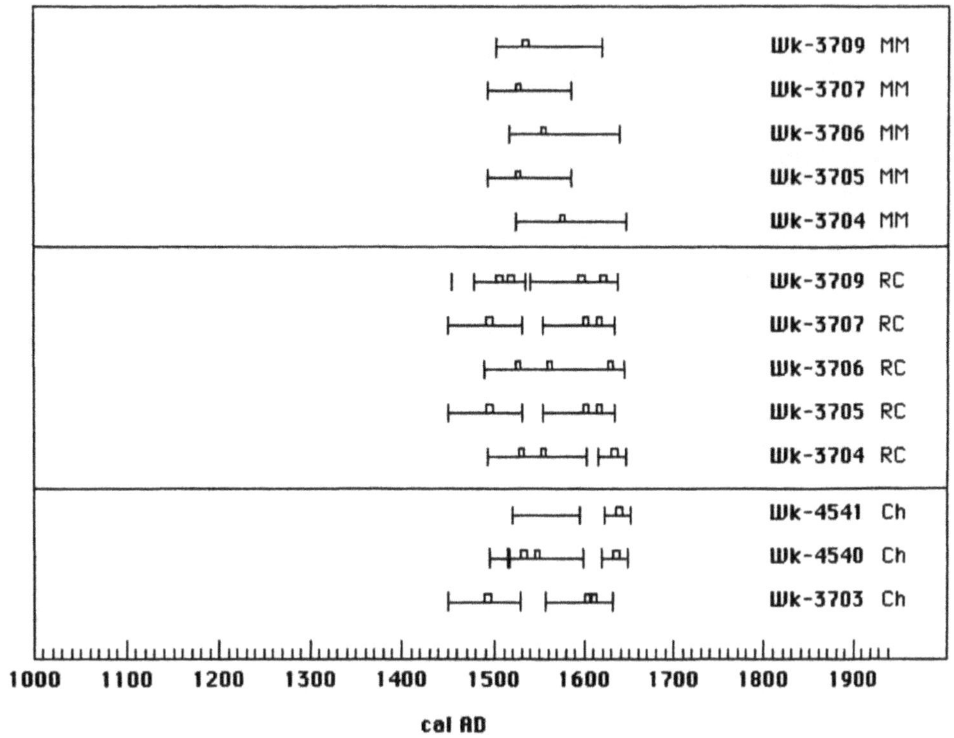


Figure 7.2. Calibrated radiocarbon ages from Kokohuia (O06/317), Layer 3. (MM = shell calibrated by Modelled Marine Curve; RC = shell calibrated using Reservoir Correction method; Ch = calibrated charcoal age).

## Layer 4

| Lab No. | Sample Type | CRA | Cal Age AD 1σ (Atmospheric Curve) | Cal Age AD 1σ (Modelled Marine Curve) |
|---|---|---|---|---|
| Wk-3711 | Charcoal | 370 ± 40 | 1491 - 1604<br>1609 - 1645 | |
| Wk-4482 | Charcoal | 450 ± 40 | 1440 - 1493<br>1603 - 1611 | |
| Wk-4483 | Charcoal | 490 ± 40 | 1430 - 1452<br>1461 - 1475 | |
| Wk-4542 | Charcoal | 430 ± 40 | 1444 - 1520<br>1592 - 1623 | |
| Wk-3712 | *Paphies australis* | 810 ± 40 | 1437 - 1489 | 1457 - 1513 |
| Wk-3713 | *Austrovenus stutchburyi* | 750 ± 40 | 1451 - 1530<br>1554 - 1633 | 1492 - 1587 |
| Wk-3715 | *Lunella smaragda* | 770 ± 40 | 1445 - 1522<br>1583 - 1625 | 1479 - 1545 |
| Wk-3716 | *Cominella virgata* | 750 ± 40 | 1451 - 1530<br>1554 - 1633 | 1492 - 1587 |

Table 7.3. Conventional radiocarbon dates and calibrated ages on charcoal and marine shell from Layer 4, Kokohuia.

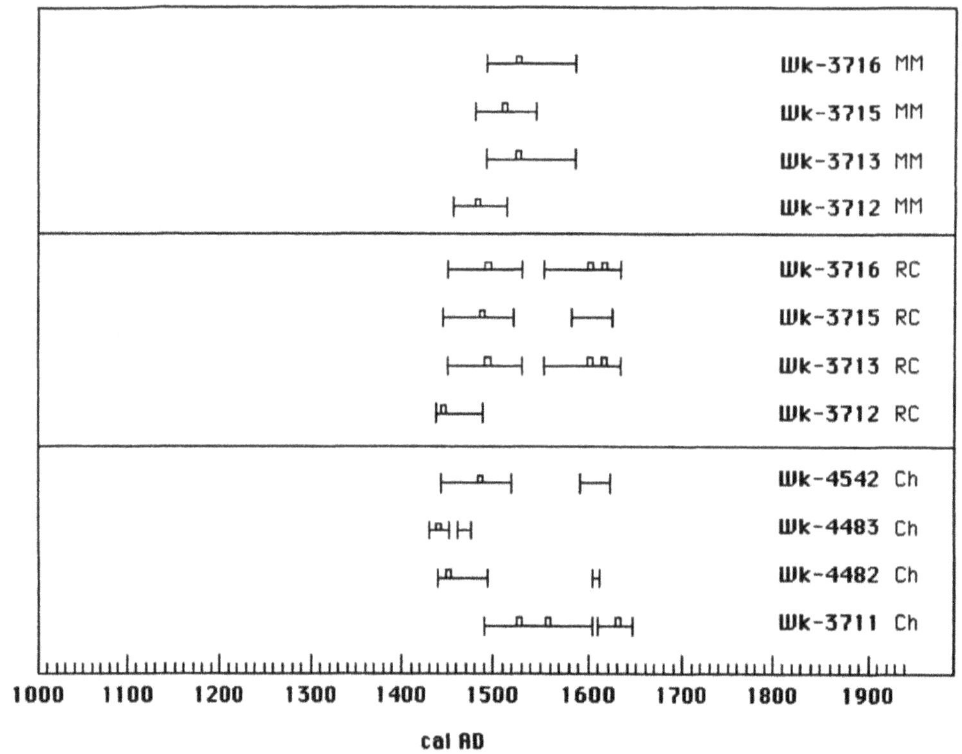


Figure 7.3. Calibrated radiocarbon ages from Kokohuia (O06/317), Layer 4. (MM = shell calibrated by Modelled Marine Curve; RC = shell calibrated using Reservoir Correction method; Ch = calibrated charcoal age).

## Layer 5

| Lab No. | Sample Type | CRA | Cal Age AD 1σ (Atmospheric Curve) | Cal Age AD 1σ (Modelled Marine Curve) |
|---|---|---|---|---|
| Wk-3718 | Charcoal | 380 ± 40 | 1487 - 1642 | |
| Wk-3719 | Charcoal | 440 ± 40 | 1442 - 1501 | |
| | | | 1510 - 1516 | |
| | | | 1599 - 1618 | |
| Wk-3720 | Charcoal | 490 ± 40 | 1430 - 1452 | |
| | | | 1461 - 1475 | |
| Wk-3721 | Charcoal | 510 ± 45 | 1419 - 1447 | |
| Wk-4543 | Charcoal | 480 ± 40 | 1434 - 1483 | |
| Wk-4544 | Charcoal | 410 ± 40 | 1450 - 1528 | |
| | | | 1558 - 1631 | |
| Wk-3722 | *Paphies australis* | 790 ± 40 | 1441 - 1495 | 1467 - 1528 |
| | | | 1601 - 1616 | |
| Wk-3723 | *Austrovenus stutchburyi* | 810 ± 40 | 1437 - 1489 | 1457 - 1513 |
| Wk-3724 | *Lunella smaragda* | 820 ± 40 | 1435 - 1486 | 1451 - 1506 |
| Wk-3726 | *Cominella virgata* | 840 ± 45 | 1427 - 1452 | 1439 - 1494 |
| | | | 1462 - 1474 | |
| Wk-3727 | *Perna canaliculus* | 820 ± 45 | 1434 - 1487 | 1449 - 1509 |

Table 7.4. Conventional radiocarbon dates and calibrated ages on charcoal and marine shell from Layer 5, Kokohuia.

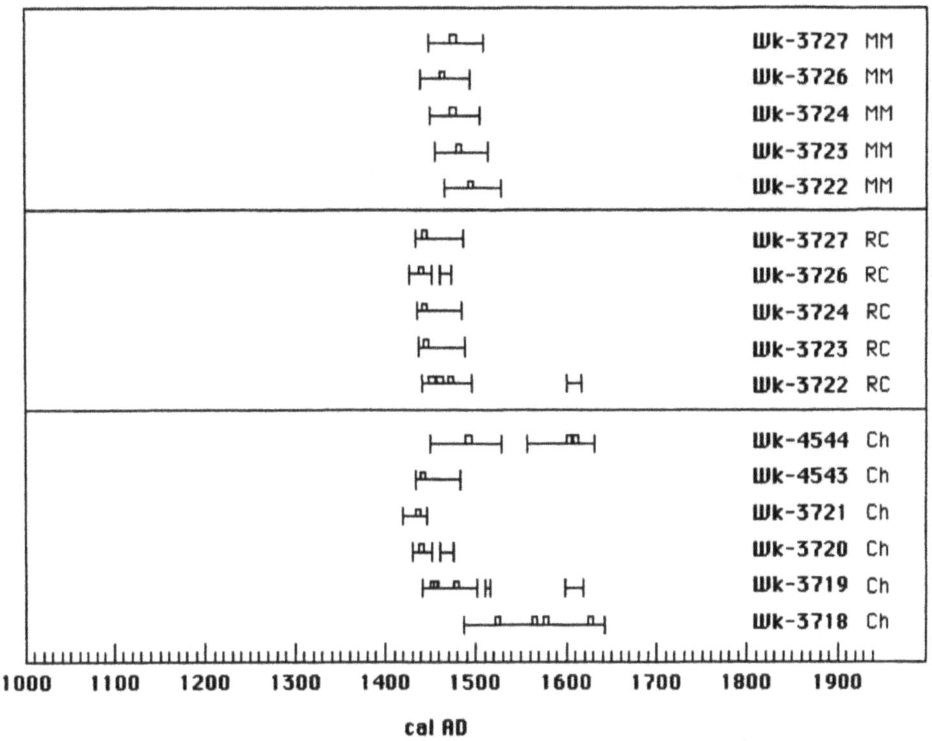


Figure 7.4. Calibrated radiocarbon ages from Kokohuia (O06/317), Layer 5. (MM = shell calibrated by Modelled Marine Curve; RC = shell calibrated using Reservoir Correction method; Ch = calibrated charcoal age).

## Discussion of Radiocarbon Dates from Kokohuia

For each cultural layer at Kokohuia, all shellfish species radiocarbon dated are in statistical agreement (Table 7.5). This suggests that the shellfish species selected was not an important variable in the accuracy of radiocarbon results. All charcoal radiocarbon dates from each layer at the site are also statistically indistinguishable (Table 7.5)

The charcoal CRAs, however, do show variability in age between the youngest and oldest dates from Layers 4 and 5. In Layer 5, for example, the minimum and maximum charcoal radiocarbon ages differ by 130 years (380 ± 40 years BP (Wk-3718) and 510 ± 45 years BP (Wk-3721) respectively). In contrast, the marine shell dates from all the layers show consistency in age. This observation is important, because if the prehistoric event represented by Layer 5 was dated using only a single charcoal sample, the time at which the event may have occurred would vary markedly depending on which sample was chosen for dating. The differences in the four charcoal ages may be statistical in nature, but the tendency for replicated charcoal results to be more variable suggests that inbuilt age may also be playing a part. Anderson (1991) has shown that even when twigs of short-lived species are chosen for dating, inbuilt age may still occur and cause discrepancies in charcoal ages from the same deposit (see Chapters 3 and 4). The environment may also cause slow decay of twigs, making ages too old. It is therefore possible the variability in charcoal ages from Layers 4 and 5 is as much to do with inbuilt age, as statistical variation.

| Provenance | Pooled Charcoal CRA (*A(p)*)* | Test Statistic*: $T$; $\chi^2_{n-1,0.05}$ | Pooled Marine Shell CRA (*A(p)*)* | Test Statistic*: $T$; $\chi^2_{n-1,0.05}$ |
|---|---|---|---|---|
| Layer 2 | 370 ± 18 | 2.48; 9.49 | 744 ± 18 | 5.83; 9.49 |
| Layer 3 | 373 ± 23 | 1.29; 5.99 | 734 ± 18 | 0.83; 9.49 |
| Layer 4 | 435 ± 20 | 4.69; 7.81 | 770 ± 20 | 1.50; 7.81 |
| Layer 5 | 450 ± 17 | 7.46; 11.07 | 815 ± 19 | 0.74; 9.49 |

Table 7.5. Pooled mean charcoal and marine shell radiocarbon ages from each cultural layer at Kokohuia. * = Ward and Wilson (1978).

| Provenance | Pooled Charcoal CRA $(A(p))$* | Test Statistic*: $T;\chi^2_{3:0.05}$ | Pooled Marine Shell CRA $(A(p))$* | Test Statistic*: $T;\chi^2_{3:0.05}$ |
|---|---|---|---|---|
| Layers 2 to 5 | 399 ± 12 | 9.32; 7.81 | 764 ± 14 | 4.59; 7.81 |

Table 7.6. Pooled charcoal and marine shell dates from layers 2 to 5 at Kokohuia have been statistically combined to obtain an overall age for the site (error in calibration curve incorporated in calculations as required in a Case 2 situation). The Test Statistics show that the pooled charcoal dates from each layer are as a group statistically distinguishable, implying that not all the cultural layers were deposited at the same time. * = Ward and Wilson (1978).

| Provenance | Pooled Charcoal CRA $(A(p))$* | $\Lambda^{\#}$ |
|---|---|---|
| Layer 2 | 370 ± 18 | 3.8 |
| Layer 3 | 373 ± 23 | 9.1 |
| Layer 4 | 435 ± 20 | 5.2 |
| Layer 5 | 450 ± 17 | |

Table 7.7. Cluster analysis on pooled mean charcoal radiocarbon dates for each layer of the Kokohuia site (calibration curve error taken into account in cluster analysis). The drop in the Λ value in Layer 4 for the charcoal chronology indicates that the pooled charcoal dates from Layers 2 and 3 can be grouped separately from the pooled charcoal ages in Layers 4 and 5. * = Ward and Wilson (1978); # = Wilson and Ward (1981).

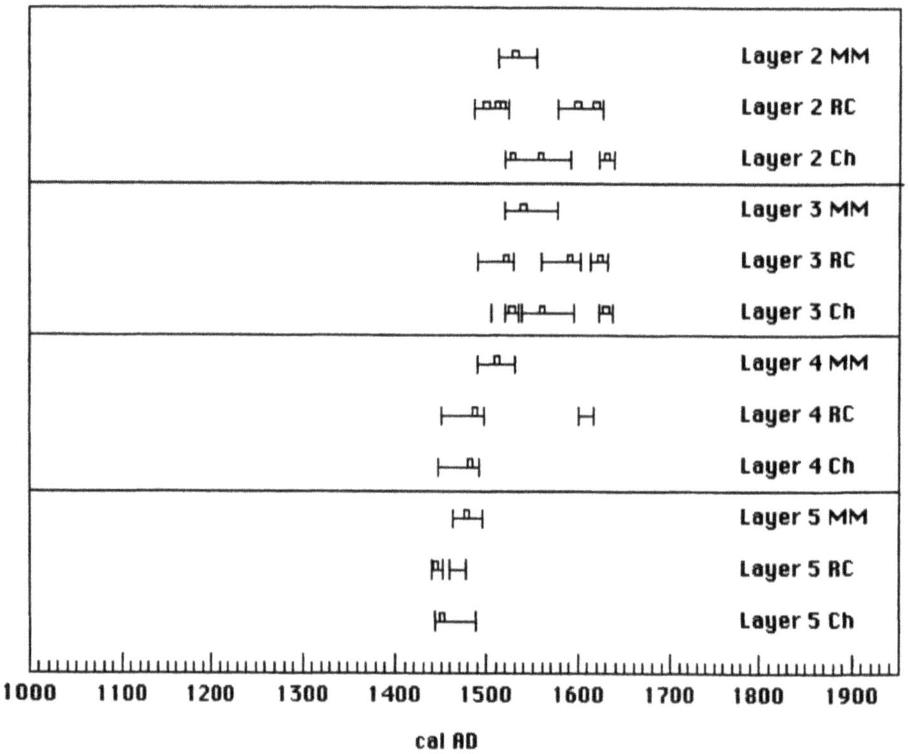


Figure 7.5. Calibrated pooled mean charcoal and marine shell radiocarbon dates from Layers 2 to 5 at Kokohuia. (MM = shell calibrated by Modelled Marine Curve; RC = shell calibrated using Reservoir Correction method; Ch = calibrated charcoal age).

The variability in the charcoal chronology at Kokohuia, affects the interpretation of prehistoric activity at the site when the history of occupation is considered. When the pooled mean charcoal and marine shell dates for all layers of the site are statistically combined (as a Case 2 situation, see Chapter 2), the charcoal ages are found to be distinguishable (Table 7.6). The charcoal chronology therefore indicates that not all the cultural layers from the site were deposited at the same time (but does not rule out continual occupation over a certain number of years), whereas the marine shell chronology implies no break in activity.

By undertaking a cluster analysis on the pooled mean charcoal dates from the site (following a Case 2b instance, see Wilson and Ward (1981:23-26)), temporal breaks in occupation may be determined.

The cluster analysis on the pooled mean charcoal ages identifies that two sub-groups of pooled mean ages are present (indicated by a drop in the $\Lambda$ value from 9.1 to 5.2) (Table 7.7). This implies there was a break in occupation between Layers 4 and 3 in prehistory, and hence the site was occupied at two differing periods. The calibrated pooled mean charcoal and marine shell radiocarbon dates from each layer at Kokohuia confirm the observations from the cluster analysis (Figure 7.5). Based on the archaeological evidence from the site, which indicates continued activity (see Chapter 6), together with the apparent greater reproducibility of the marine shell dates from each layer, the occupational history at Kokohuia is most probably reflected through the marine shell chronology. This, therefore, indicates that the marine shell ages from Kokohuia more closely reflect actual prehistoric behaviour at the site, than the charcoal chronology does.

The calibrated charcoal radiocarbon ages from each layer at Kokohuia are in agreement with their calibrated marine shell pairs (Figures 7.1 - 7.4). This suggests that hard water has not affected the marine shell ages from this site, despite the highly calcareous geology surrounding Kokohuia. Two differences between the two marine shell calibration methods are apparent. First, the modelled marine curve often produces an age range for each calibrated marine shell date, which is less spread over time than that produced by the reservoir correction method. Second, the modelled marine curve calibrated $^{14}C$ dates appear to be slightly younger than reservoir corrected calibrated radiocarbon dates. The spread in age ranges for the reservoir corrected calibrated shell dates, is due to calibration on the atmospheric curve which reflects temporal fluctuations in atmospheric $^{14}C$ concentration. The older reservoir corrected calibrated shell ages may be due to the use of a fixed $R$ correction which is not applicable for this period of prehistory. It can also be seen that shell CRAs greater than 800 ± 40 years BP calibrated using both marine calibration methods, are in closer agreement in terms of calibrated age range, than calibrated shell dates less than 790 ± 40 years BP. This appears to be due to the smoother nature of the atmospheric calibration curve before 1500 cal AD.

## Aotea

***Area B (S8/171), Layer 2*** *(Table 7.8; Figure 7.6)*

All marine shell species dated from Layer 2 of Area B, yielded statistically identical results ($A(p)$ = 657 ± 23 ($T$ = 0.79; $\chi^2_{2:0.05}$ = 5.99)). Only one charcoal radiocarbon age was obtained from this site because of a paucity of charcoal retrieved during excavation. When calibrated, however, this charcoal radiocarbon age is in agreement with the calibrated marine shell $^{14}C$ dates (Figure 7.6).

The calibrated shell dates from this site illustrate clearly the differences between calibrating shell ages using either the reservoir correction or modelled marine curve methods. Whereas the latter provides a chronology with less age range, the former gives age ranges which span a greater amount of calendar time (Figure 7.6). This spread in age ranges is due to 'wiggles' in the atmospheric calibration curve after *ca.* 1500 cal AD. This effect is also apparent in the calibrated charcoal radiocarbon age. Area B was believed by the excavators to have been occupied around 250 years BP (see Chapter 6). Shell dates calibrated using the modelled marine curve are more consistent with the excavator's estimate.

| Lab No. | Sample Type | CRA | Cal Age AD 1σ (Atmospheric Curve) | Cal Age AD 1σ (Modelled Marine Curve) |
|---|---|---|---|---|
| Wk-3467 | Charcoal | 220 ± 40 | 1667 - 1690<br>1730 - 1811<br>1923 - 1954 | |
| Wk-3678 | *Paphies subtriangulatum* | 660 ± 40 | 1534 - 1540<br>1637 - 1662 | 1610 - 1673 |
| Wk-3679 | *Paphies australis* | 680 ± 40 | 1525 - 1562<br>1629 - 1654 | 1549 - 1663 |
| Wk-3680 | *Austrovenus stutchburyi* | 630 ± 40 | 1646 - 1670<br>1782 - 1798<br>1947 - 1953 | 1642 - 1687 |

Table 7.8. Conventional radiocarbon dates and calibrated ages on charcoal and marine shell from Area B, Layer 2, Aotea.

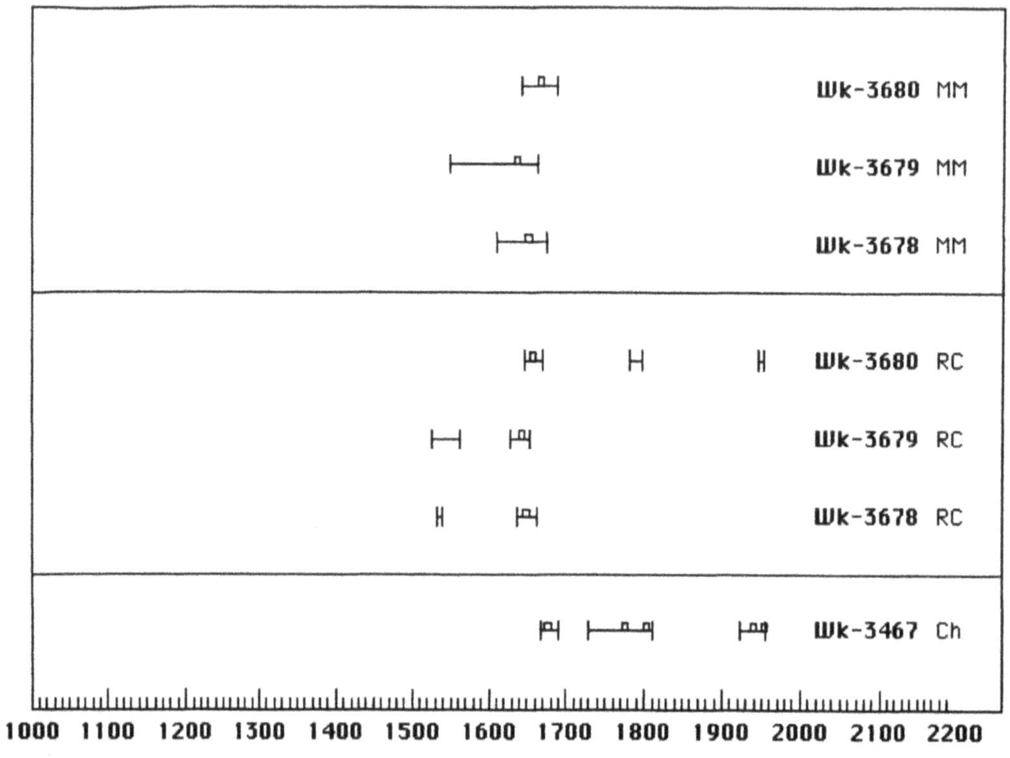


Figure 7.6. Calibrated radiocarbon ages from Site B (S8/171), Layer 2, Aotea. (MM = shell calibrated by Modelled Marine Curve; RC = shell calibrated using Reservoir Correction method; Ch = calibrated charcoal age).

## Ponui Island

The main difficulty in radiocarbon dating the archaeological sites excavated on Ponui Island, was obtaining adequate amounts of short-lived charcoal from each site. The island was dominated by ancient kauri-podocarp-hardwood forest during occupation, and so short-lived species were rarely found in secure cultural contexts (Wallace pers. comm.).

***Site N43/1, Layer D*** *(Table 7.9; Figure 7.7)*

Despite the fact that dated shellfish species came from rocky shore, estuarine and sandy beach environments, all radiocarbon ages are in statistical agreement ($A(p)$ = 843 ± 20 ($T$ = 0.55; $\chi^2_{4:0.05}$ = 7.81)). All calibrated marine shell determinations except one, are in agreement with the calibrated charcoal sample, when using either shell calibration method (Figure 7.7). The lower age range of Wk-3581 calibrated using the modelled marine curve, falls just outside the upper age range of the calibrated charcoal radiocarbon date (Table 7.9). Because the age difference is only by 0.25σ (nine radiocarbon years), and the CRA is statistically indistinguishable from other marine shell determinations from the same stratigraphic horizon, this is probably not a significant result. For this site, then, the calibrated shell ages obtained using the reservoir correction calibration method, are in closer agreement with the calibrated charcoal age, than the shell dates calibrated using the modelled marine curve, but as only one charcoal radiocarbon date was derived from this site, the significance of this result is hampered by lower levels of precision.

| Lab No. | Sample Type | CRA | Cal Age AD 1σ (Atmospheric Curve) | Cal Age AD 1σ (Modelled Marine Curve) |
|---|---|---|---|---|
| NZ-7764 | Charcoal | 535 ± 48 | 1410 - 1442 | |
| Wk-3578 | *Paphies australis* | 840 ± 40 | 1428 - 1451 | 1441 - 1491 |
| Wk-3579 | *Austrovenus stutchburyi* | 850 ± 40 | 1425 - 1447 | 1436 - 1485 |
| Wk-3580 | *Venerupis largillierti* | 860 ± 40 | 1419 - 1445 | 1431 - 1479 |
| Wk-3581 | *Perna canaliculus* | 820 ± 40 | 1435 - 1486 | 1451 - 1506 |

Table 7.9. Conventional radiocarbon dates and calibrated ages on charcoal and marine shell from N43/1, Layer D, Ponui Island.

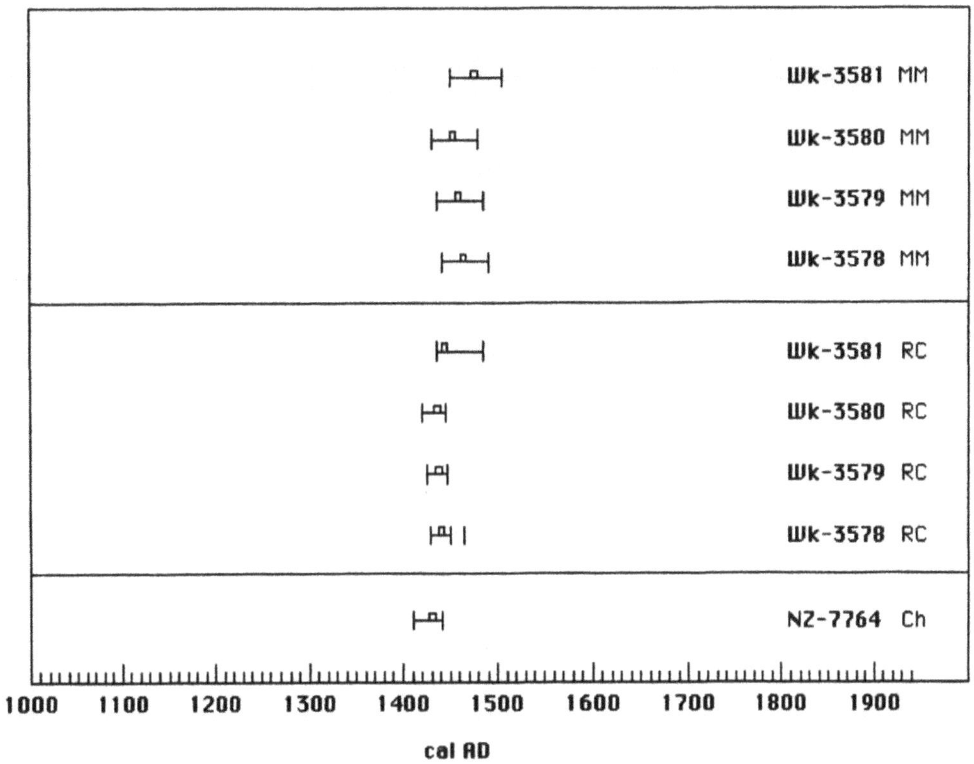


Figure 7.7. Calibrated radiocarbon ages from site N43/1, Layer D, Ponui Island. (MM = shell calibrated by Modelled Marine Curve; RC = shell calibrated using Reservoir Correction method; Ch = calibrated charcoal age).

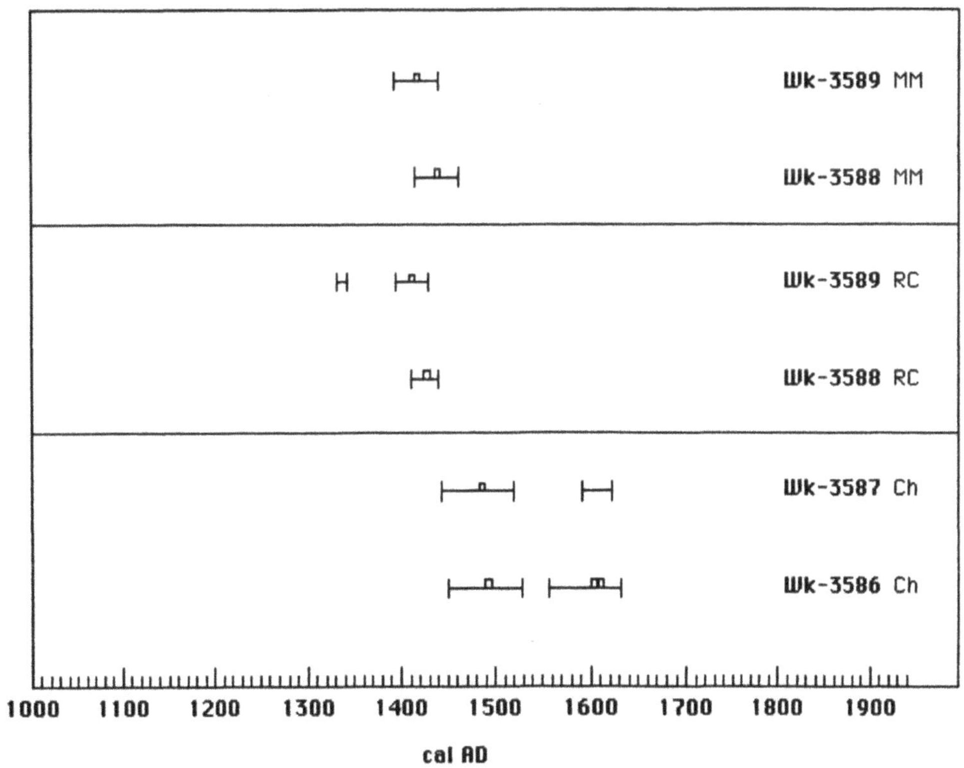


Figure 7.8. Calibrated radiocarbon ages from site N43/333, Layer D, Ponui Island. (MM = shell calibrated by Modelled Marine Curve; RC = shell calibrated using Reservoir Correction method; Ch = calibrated charcoal age).

| Lab No. | Sample Type | CRA | Cal Age AD 1σ (Atmospheric Curve) | Cal Age AD 1σ (Modelled Marine Curve) |
|---|---|---|---|---|
| Wk-3586 | Charcoal | 410 ± 40 | 1449 - 1528<br>1558 - 1631 | |
| Wk-3587 | Charcoal | 430 ± 40 | 1443 - 1520<br>1591 - 1623 | |
| Wk-3588 | *Paphies australis* | 890 ± 40 | 1410 - 1439 | 1415 - 1461 |
| Wk-3589 | *Austrovenus stutchburyi* | 930 ± 40 | 1332 - 1342<br>1394 - 1429 | 1392 - 1441 |

Table 7.10. Conventional radiocarbon dates and calibrated ages on charcoal and marine shell from N43/333, Layer D, Ponui Island.

| Lab No. | Sample Type | CRA | Cal Age AD 1σ (Atmospheric Curve) | Cal Age AD 1σ (Modelled Marine Curve) |
|---|---|---|---|---|
| Wk-3590 | Charcoal | 620 ± 40 | 1325 - 1353<br>1360 - 1367<br>1388 - 1413 | |
| Wk-3591 | *Paphies australis* | 920 ± 45 | 1333 - 1339<br>1397 - 1434 | 1395 - 1448 |
| Wk-3592 | *Austrovenus stutchburyi* | 940 ± 40 | 1330 - 1346<br>1392 - 1425 | 1381 - 1436 |
| Wk-3593 | *Perna canaliculus* | 910 ± 40 | 1406 - 1436 | 1403 - 1450 |
| Wk-3594 | *Crassostrea glomerata* | 900 ± 40 | 1408 - 1437 | 1409 - 1456 |

Table 7.11. Conventional radiocarbon dates and calibrated ages on charcoal and marine shell from Site 14, Layer D, Ponui Island.

| Lab No. | Sample Type | CRA | Cal Age AD 1σ (Atmospheric Curve) | Cal Age AD 1σ (Modelled Marine Curve) |
|---|---|---|---|---|
| Wk-3596 | Charcoal | 470 ± 40 | 1436 - 1487 | |
| Wk-3597 | *Paphies australis* | 740 ± 40 | 1477 - 1534<br>1538 - 1637 | 1506 - 1605 |
| Wk-3598 | *Austrovenus stutchburyi* | 820 ± 40 | 1435 - 1486 | 1450 - 1507 |

Table 7.12. Conventional radiocarbon dates and calibrated ages on charcoal and marine shell from Site 27, Layer D, Ponui Island.

***Site N43/333, Layer D** (Table 7.10; Figure 7.8)*

The two shellfish species dated from this site are statistically indistinguishable ($A(p)$ = 910 ± 28 ($T$ = 0.50; $\chi^2_{1:0.05}$ = 3.84)). The charcoal CRAs are also in statistical agreement ($A(p)$ = 420 ± 28 ($T$ = 0.13; $\chi^2_{1:0.05}$ = 3.84)). Age ranges from the modelled marine curve calibrated shell dates and calibrated charcoal radiocarbon dates from this layer show close agreement, and appear to correctly date this deposit as evidenced by the presence of an Archaic Duff 1A adze found in the site (see Chapter 6) (Figure 7.8). The age ranges of the reservoir corrected calibrated shell dates, however, fall outside the age ranges of the calibrated charcoal date. Although one of the modelled marine curve calibrated dates upper age range (Wk-3589) falls just outside the minimum age range of the oldest calibrated charcoal radiocarbon date (Wk-3587), this is only by two radiocarbon years. In addition, Wk-3589 is statistically indistinguishable from its partnered marine shell age. Marine shell radiocarbon ages from different shellfish species at this site are accurate, then, but in terms of calibration, the modelled marine curve calibration method produces the closest age ranges to those obtained from the calibrated charcoal dates.

***Site 14, Layer D** (Table 7.11; Figure 7.9)*

Radiocarbon age determinations on marine shell from Site 14 are statistically indistinguishable ($A(p)$ = 917 ± 21 ($T$ = 0.55; $\chi^2_{3:0.05}$ = 7.81)). All marine shell CRAs calibrated using both marine shell calibration methods are in close agreement with the paired calibrated charcoal date. Because of fluctuations in the atmospheric calibration curve, though, marine shell radiocarbon dates Wk-3591 and Wk-3592 calibrated using the reservoir correction method, have more extensive age ranges in comparison to those obtained using the modelled marine curve (Figure 7.9). The calibrated charcoal age also experiences multiple age ranges. Both marine shell calibration methods, therefore, are in close agreement in dating this archaeological deposit.

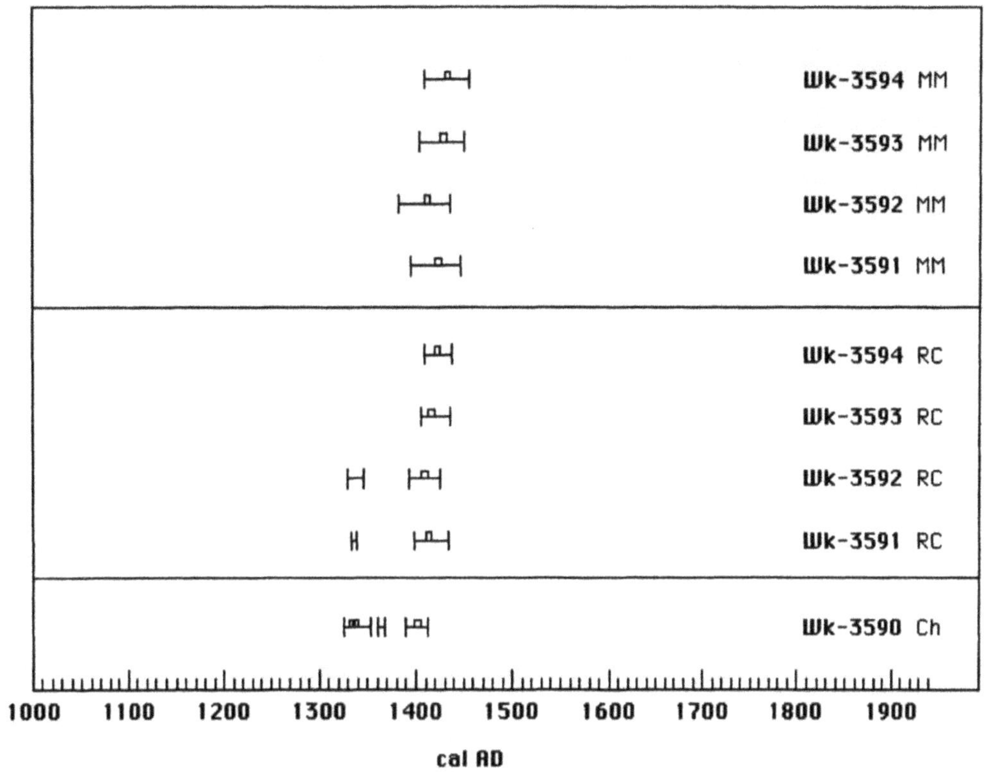


Figure 7.9. Calibrated radiocarbon ages from Site 14, Ponui Island. (MM = shell calibrated by Modelled Marine Curve; RC = shell calibrated using Reservoir Correction method; Ch = calibrated charcoal age).

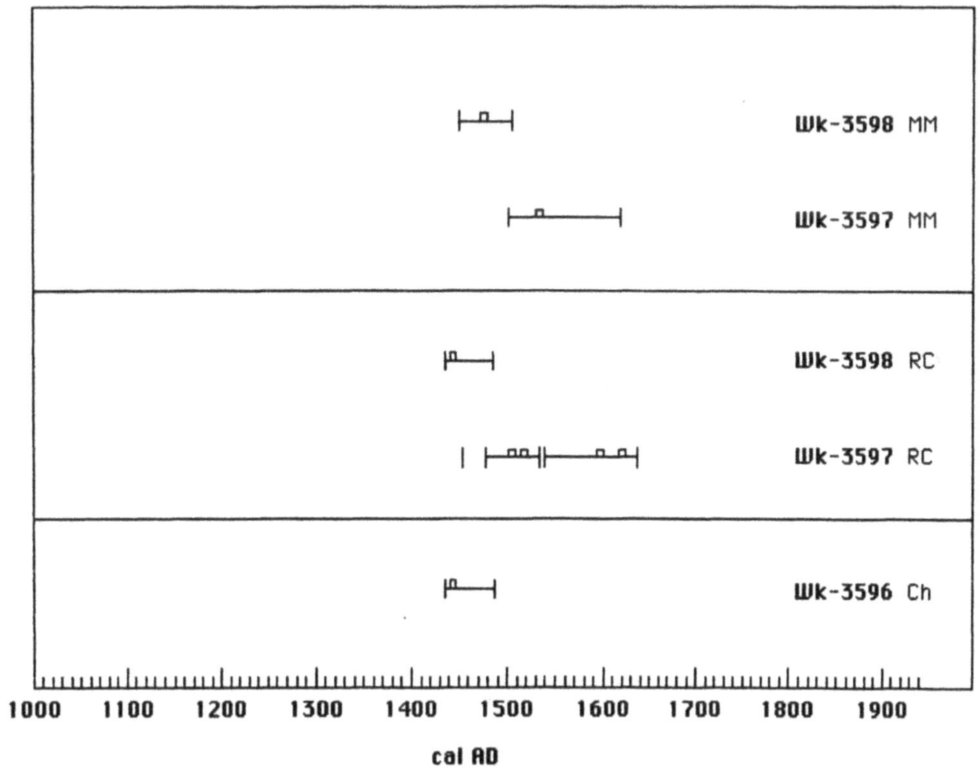


Figure 7.10. Calibrated radiocarbon ages from Site 27, Ponui Island. (MM = shell calibrated by Modelled Marine Curve; RC = shell calibrated using Reservoir Correction method; Ch = calibrated charcoal age).

***Site 27, Layer D*** *(Table 7.12; Figure 7.10)*

Both marine shell species are in agreement for dating this prehistoric event ($A(p)$ = 780 ± 28 ($T$ = 2.00; $\chi^2_{3:0.05}$ = 3.84)). When the shell dates are calibrated using either marine shell calibration method, Wk-3597 calibrated using the modelled marine curve is offset compared to its paired charcoal calibrated age (Wk-3596) (Table 7.12; Figure 7.10). Again, however, the difference is small, by only 0.5 σ (19 years). For this site, therefore, the reservoir corrected calibrated shell dates are in closer agreement with the calibrated charcoal age, than shell ages calibrated using the modelled marine curve.

## Ligar Bay

***Site N25/95, Middens 1 And 2*** *(Tables 7.13 - 7.14, 7.22; Figures 7.11 - 7.12)*

Because both of these middens were close to one another and stratigraphically related, they are assumed to be contemporary and are discussed together here. Both middens show close agreement between radiocarbon determinations from all shellfish species ($A(p)$ = 856 ± 14 ($T$ = 7.62; $\chi^2_{7:0.05}$ = 14.07)). Fast decay rates of dead wood and short life spans of trees in the area, may have been an influence on the close similarity between dates of charcoal from the site ($A(p)$ = 455 ± 28 ($T$ = 0.28; $\chi^2_{1:0.05}$ = 3.84)) (see Chapter 4). Certainly, short-lived charcoal was easier to obtain from the Ligar Bay sites than it was at Kokohuia and Ponui Island.

For both middens, the modelled marine curve calibration method produces calibrated age ranges which are in closer agreement with the age ranges from the calibrated charcoal dates, than the reservoir corrected calibrated shell dates.

The upper calibrated age range of sample Wk-3546 from Midden 2 calibrated using the reservoir correction method, falls just outside the calibrated age range of its paired charcoal age Wk-3544 (Table 7.14; Figure 7.12). Again, this difference is minimally significant (three radiocarbon years separating the upper age range of Wk-3546, from the lower age range of the calibrated charcoal $^{14}C$ age (Wk-3544)). Also, Wk-3546 is statistically indistinguishable from other marine shell determinations from the same horizon. Wk-3545 and Wk-3546 from midden 2 exhibit extended age ranges when calibrated using the reservoir correction method. Again, the reason is due to 'wiggles' in the atmospheric calibration curve at *ca.* 920 years BP, and 790 years BP.

***N25/26, Layer 4*** *(Table 7.15, Figure 7.13)*

Radiocarbon determinations obtained from Layer 4 of site N25/26 show that the species of shellfish dated was not a variable of importance in affecting reproducibility: all results are statistically indistinguishable regardless of species ($A(p)$ = 742 ± 18 ($T$ = 1.55; $\chi^2_{4:0.05}$ = 9.49)). Both marine shell radiocarbon age calibration methods produce results in close agreement with the paired charcoal calibrated age. The only difference between both marine shell calibration methods, is that shell dates calibrated using the modelled marine curve are less spread in age ranges in comparison to those calibrated using the reservoir correction method, and produce a better 'fit' with the calibrated charcoal date (Figure 7.13). The close agreement between the charcoal and marine shell determinations is probably due to the deliberate selection of charcoal identified to twigs of short-lived species prior to dating, and the high rate of decay of wood in the local environment (see Chapter 6).

| Lab No. | Sample Type | CRA | Cal Age AD 1σ (Atmospheric Curve) | Cal Age AD 1σ (Modelled Marine Curve) |
|---|---|---|---|---|
| Wk-3539 | Charcoal | 440 ± 40 | 1441 - 1501<br>1510 - 1516<br>1598 - 1618 | |
| Wk-3540 | *Paphies subtriangulatum* | 880 ± 40 | 1412 - 1441 | 1420 - 1467 |
| Wk-3541 | *Paphies australis* | 820 ± 40 | 1435 - 1486 | 1451 - 1506 |
| Wk-3542 | *Austrovenus stutchburyi* | 840 ± 40 | 1428 - 1451 | 1441 - 1491 |
| Wk-3543 | *Lunella smaragda* | 870 ± 40 | 1414 - 1443 | 1440 - 1493 |

Table 7.13. Conventional radiocarbon dates and calibrated ages on charcoal and marine shell from Midden 1, site N25/95, Ligar Bay.

| Lab No. | Sample Type | CRA | Cal Age AD 1σ (Atmospheric Curve) | Cal Age AD 1σ (Modelled Marine Curve) |
|---|---|---|---|---|
| Wk-3544 | Charcoal | 470 ± 40 | 1436 - 1487 | |
| Wk 3545 | *Paphies subtriangulatum* | 790 ± 40 | 1440 - 1495<br>1601 - 1616 | 1467 - 1528 |
| Wk 3546 | *Paphies australis* | 920 ± 40 | 1334 - 1337<br>1401 - 1433 | 1398 - 1445 |
| Wk 3547 | *Austrovenus stutchburyi* | 840 ± 40 | 1428 - 1451 | 1441 - 1491 |
| Wk 3548 | *Lunella smaragda* | 890 ± 40 | 1410 - 1439 | 1415 - 1461 |

Table 7.14. Conventional radiocarbon dates and calibrated ages on charcoal and marine shell from Midden 2, site N25/95, Ligar Bay.

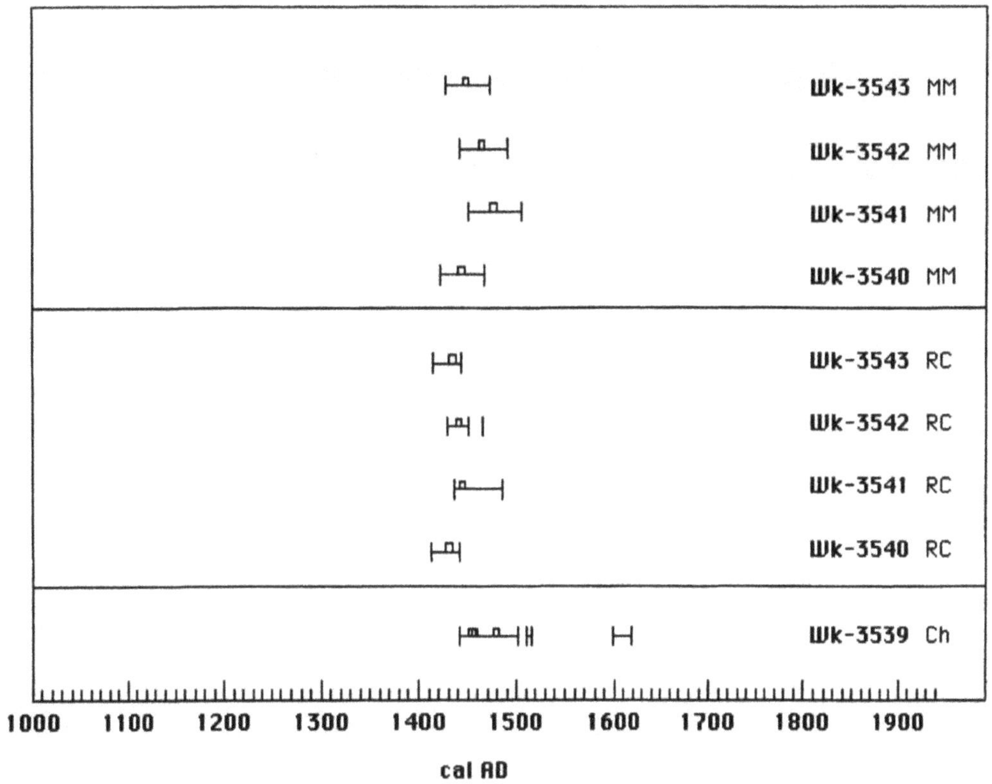


Figure 7.11. Calibrated radiocarbon ages from site N25/95, Midden 1, Ligar Bay. (MM = shell calibrated by Modelled Marine Curve; RC = shell calibrated using Reservoir Correction method; Ch = calibrated charcoal age).

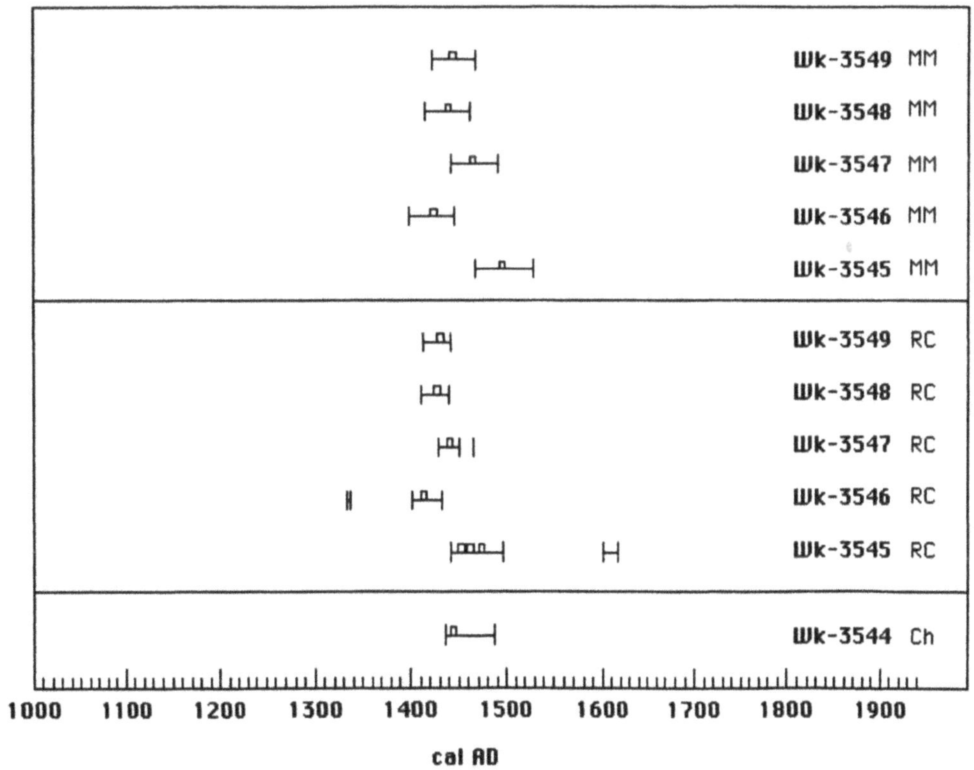


Figure 7.12. Calibrated radiocarbon ages from site N25/95, Midden 2, Ligar Bay. (MM = shell calibrated by Modelled Marine Curve; RC = shell calibrated using Reservoir Correction method; Ch = calibrated charcoal age).

| Lab No. | Sample Type | CRA | Cal Age AD 1σ (Atmospheric Curve) | Cal Age AD 1σ (Modelled Marine Curve) |
|---|---|---|---|---|
| Wk-3533 | Charcoal | 390 ± 40 | 1482 - 1638 | |
| Wk-3534 | *Paphies subtriangulatum* | 740 ± 40 | 1477 - 1534<br>1538 - 1637 | 1500 - 1619 |
| Wk-3535 | *Paphies australis* | 760 ± 40 | 1448 - 1526<br>1561 - 1629 | 1485 - 1560 |
| Wk-3536 | *Austrovenus stutchburyi* | 750 ± 40 | 1451 - 1530<br>1554 - 1633 | 1492 - 1587 |
| Wk-3537 | *Lunella smaragda* | 760 ± 40 | 1448 - 1526<br>1561 - 1629 | 1485 - 1560 |
| Wk-3538 | *Cominella adspersa* | 700 ± 40 | 1501 - 1507<br>1517 - 1597<br>1620 - 1649 | 1530 - 1653 |

Table 7.15. Conventional radiocarbon dates and calibrated ages on charcoal and marine shell from site N25/26, Layer 4, Ligar Bay.

| Lab No. | Sample Type | Layer | CRA | Cal Age AD 1σ (Atmospheric Curve) | Cal Age AD 1σ (Modelled Marine Curve) |
|---|---|---|---|---|---|
| Wk-4028 | Charcoal | L1B | 350 ± 40 | 1519 - 1594<br>1622 - 1650 | |
| Wk-4033 | *Paphies australis* | L1B | 830 ± 40 | 1432 - 1454<br>1458 - 1479 | 1446 - 1498 |
| Wk-4030 | *Austrovenus stutchburyi* | L1B | 790 ± 40 | 1441 - 1495<br>1601 - 1616 | 1467 - 1528 |
| NZ-7701i | Charcoal | L1C | 310 ± 110 | 1447 - 1671<br>1781 - 1799<br>1945 - 1953 | |
| Wk-4029 | Charcoal | L1C | 410 ± 40 | 1450 - 1528<br>1558 - 1631 | |
| Wk-3099 | *Paphies australis* | L1C | 800 ± 50 | 1437 - 1495<br>1601 - 1616 | 1457 - 1527 |
| Wk-4031 | *Lunella smaragda* | L1C | 800 ± 40 | 1439 - 1492 | 1462 - 1520 |

Table 7.16. Conventional radiocarbon dates and calibrated ages on charcoal and marine shell from Layers L1B and L1C, Sawpit Point (N26/214), Awaroa.

## Awaroa

***Sawpit Point (N26/214)** (Table 7.16; Figure 7.14)*

Radiocarbon determinations from Layers L1B and L1C confirm Barber's (pers. comm.) suggestion that the layers are contemporary (see Chapter 6). All marine shell radiocarbon ages are statistically indistinguishable ($A(p)$ = 805 ± 21 ($T$ = 0.56; $\chi^2_{3:0.05}$ = 7.81)), as are the charcoal dates ($A(p)$ = 376 ± 27 ($T$ = 1.50; $\chi^2_{2:0.05}$ = 5.99)).

The reservoir corrected calibrated marine shell $^{14}C$ dates are in agreement with the modelled marine calibrated radiocarbon dates, but have greater age ranges, and are slightly older than the calibrated charcoal dates, and modelled marine curve calibrated shell dates. The closeness of the calibrated charcoal and shell dates, suggests that there is limited inbuilt age in the charcoal from the site, as found at the Ligar Bay sites on the same coastline.

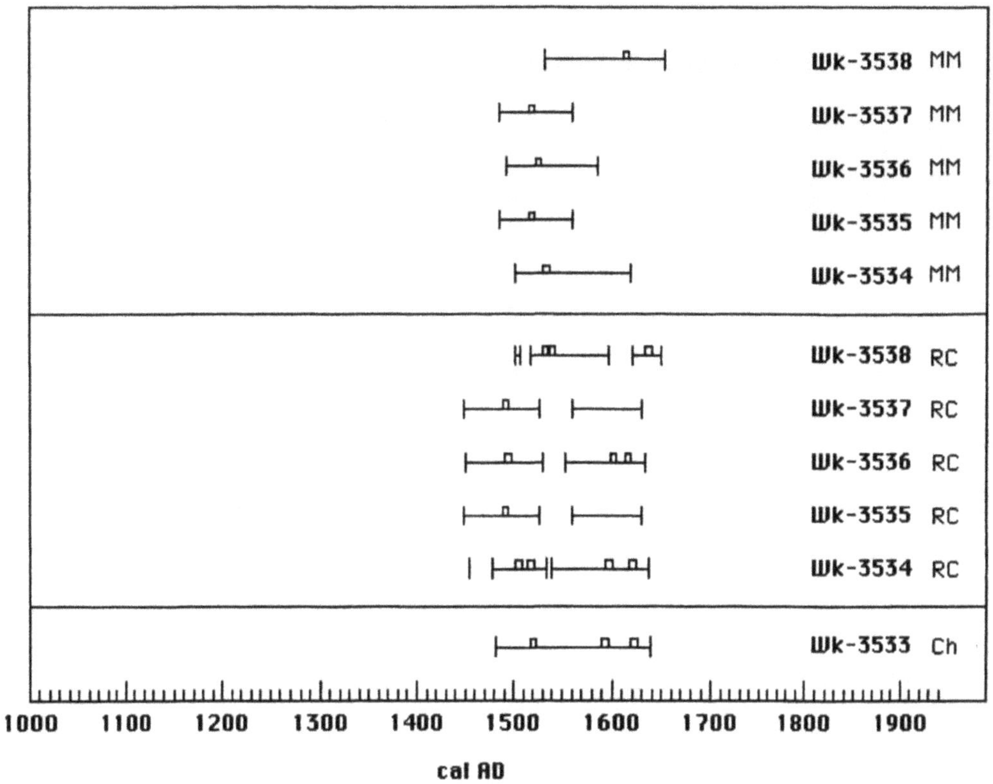


Figure 7.13. Calibrated radiocarbon ages from site N25/26, Ligar Bay. (MM = shell calibrated by Modelled Marine Curve; RC = shell calibrated using Reservoir Correction method; Ch = calibrated charcoal age).

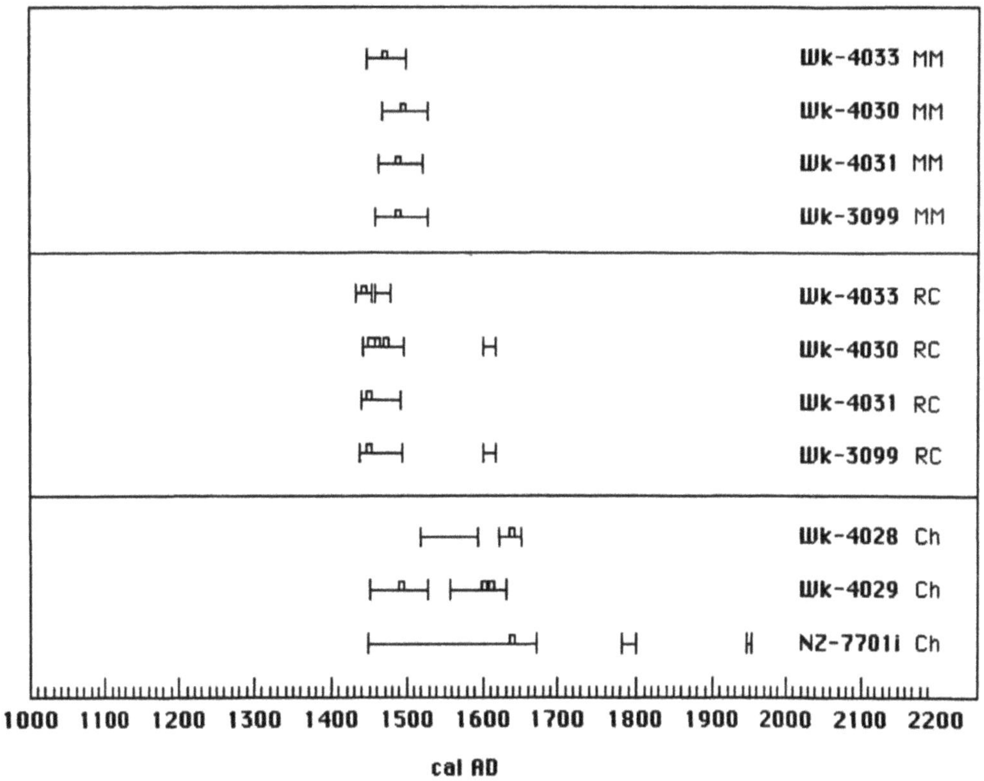


Figure 7.14. Calibrated radiocarbon ages from Sawpit Point (N26/214), Layers L1B and L1C, Awaroa. (MM = shell calibrated by Modelled Marine Curve; RC = shell calibrated using Reservoir Correction method; Ch = calibrated charcoal age).

## ARCHAEOLOGICAL SITES WITH NO CHARCOAL/MARINE SHELL CHRONOLOGICAL AGREEMENT

Five archaeological sites yielded calibrated charcoal and marine shell radiocarbon ages which were in disagreement, regardless of which marine shell calibration method was used (Tables 7.17 - 7.21; Figures 7.15 - 7.20). The most likely reasons for the observed differences in radiocarbon age between these sample types for each of the affected sites are discussed below.

### Inbuilt Age

#### *Site N43/333, Layer B, Ponui Island*

Layer B of site N43/333 shows that the calibrated charcoal radiocarbon age from this archaeological deposit is significantly older than the calibrated marine shell radiocarbon ages from the identical cultural layer (Table 7.17; Figure 7.15). Layer D (stratigraphically below Layer B) of site N43/333 (see Table 7.10 and Figure 7.8), showed agreement between its calibrated marine shell and charcoal radiocarbon ages, and indicated that the site was occupied *ca.* 400 years BP. However, Layer B is well separated from Layer D and near the surface of the site implying more recent deposition. Given the closeness of the marine shell and charcoal radiocarbon ages from Layer D of this site, the charcoal/marine shell paired radiocarbon dates from other archaeological sites on the island, and from other locations around the country, inbuilt age is probably one explanation for the anomalous charcoal age from Layer B. It was illustrated at the Kokohuia site that radiocarbon dates on twig charcoal from the same deposit can vary markedly, possibly due to inbuilt age. It appears, then, that the charcoal dated from Layer B may have contained either twig wood which had survived in the environment for a great period of time, or mis-identified twig material, which actually contained branch wood representing many years of growth, or long dead heart wood.

| Lab No. | Sample Type | CRA | Cal Age AD 1σ (Atmospheric Curve) | Cal Age AD 1σ (Modelled Marine Curve) |
|---|---|---|---|---|
| Wk-3582 | Charcoal | 400 ± 40 | 1453 - 1461<br>1471 - 1532<br>1546 - 1635 | |
| Wk-3583 | *Paphies australis* | 590 ± 50 | 1654 - 1684<br>1733 - 1808<br>1931 - 1954 | 1660 - 1719 |
| Wk-3584 | *Austrovenus stutchburyi* | 520 ± 40 | 1678 - 1775<br>1804 - 1891<br>1909 - 1939<br>1954 - 1955* | 1706 - 1827 |
| Wk-3585 | *Crassostrea glomerata* | 520 ± 40 | 1678 - 1775<br>1804 - 1891<br>1909 - 1939<br>1954 - 1955* | 1706 - 1827 |

Table 7.17. Conventional radiocarbon dates and calibrated ages on charcoal and marine shell from site N43/333, Layer B, Ponui Island. (* denotes influence of bomb C-14, see Stuiver and Reimer (1993)).

| Lab No. | Sample Type | CRA | Cal Age AD 1σ (Atmospheric Curve) | Cal Age AD 1σ (Modelled Marine Curve) |
|---|---|---|---|---|
| Wk-3463 | Charcoal | MODERN^ | MODERN | |
| Wk-3464 | *Paphies subtriangulatum* | 720 ± 60 | 1479 - 1649 | 1501 - 1652 |
| Wk-3465 | *Paphies australis* | 640 ± 30 | 1643 - 1668<br>1784 - 1796<br>1950 - 1952 | 1642 - 1678 |
| Wk-3466-1 | *Austrovenus stutchburyi* | 510 ± 40 | 1681 - 1739<br>1806 - 1901<br>1906 - 1936<br>1954 - 1955* | 1705 - 1854 |
| Wk-3466-2 | *Austrovenus stutchburyi* | 570 ± 40 | 1665 - 1688<br>1731 - 1810<br>1925 - 1954 | 1675 - 1735 |

Table 7.18. Conventional radiocarbon dates and calibrated ages on charcoal and marine shell from Area A, Layer 2, Aotea. (^ though Wk-3463 is MODERN, the CRA of 198 ± 40 has been calibrated in figure 7.16 to illustrate the differences between calibrating young charcoal CRA using the atmospheric calibration curve, and calibrating young shell CRA using the modelled marine curve.)

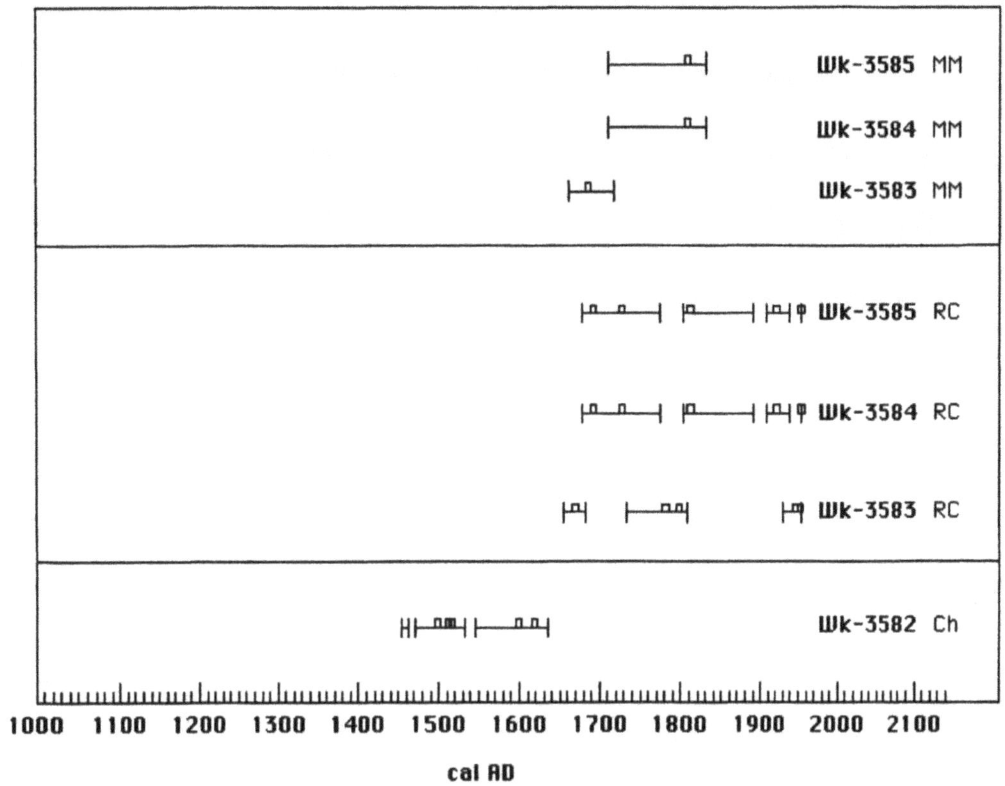


Figure 7.15. Calibrated radiocarbon ages from site N43/333, Layer B, Ponui Island. (MM = shell calibrated by Modelled Marine Curve; RC = shell calibrated using Reservoir Correction method; Ch = calibrated charcoal age).

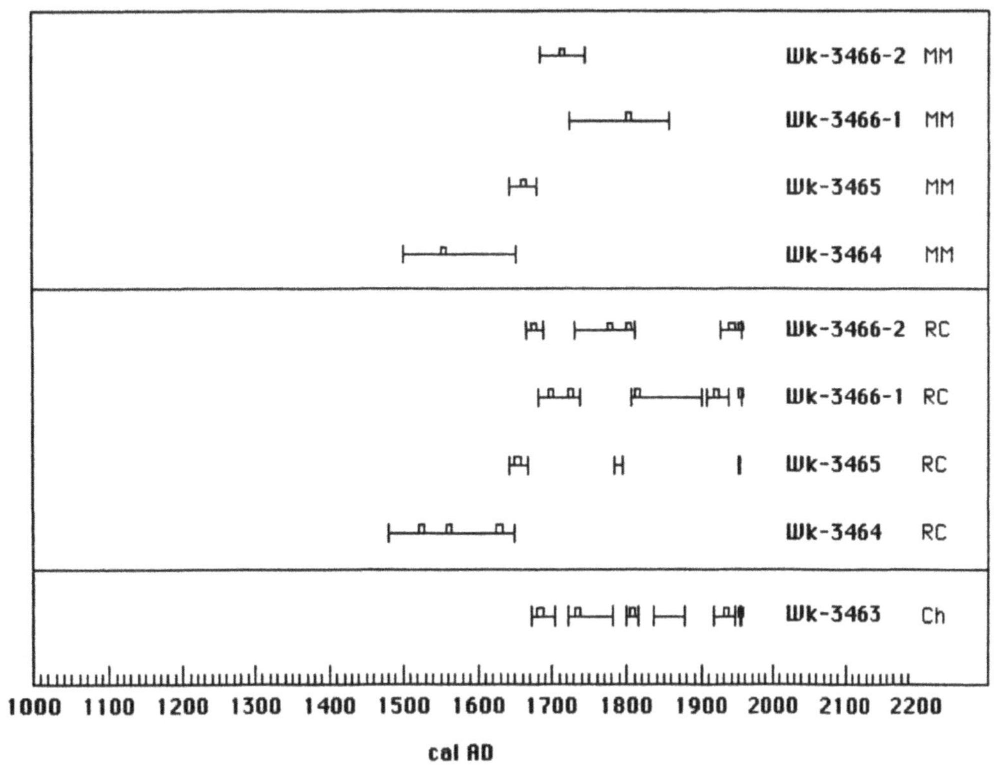


Figure 7.16. Calibrated radiocarbon ages from Area A (S8/171), Layer 2, Aotea. (MM = shell calibrated by Modelled Marine Curve; RC = shell calibrated using Reservoir Correction method; Ch = calibrated charcoal age).

| Lab No. | Sample Type | CRA | Cal Age AD 1σ (Atmospheric Curve) | Cal Age AD 1σ (Modelled Marine Curve) |
|---|---|---|---|---|
| Wk-3468 | Charcoal | 220 ± 40 | 1667 - 1690<br>1730 - 1811<br>1923 - 1954 | |
| Wk-3471 | *Paphies subtriangulatum* | 730 ± 50 | 1479 - 1643 | 1501 - 1639 |
| Wk-3470 | *Paphies australis* | 730 ± 40 | 1486 - 1640 | 1508 - 1632 |
| Wk-3469 | *Austrovenus stutchburyi* | 820 ± 40 | 1435 - 1486 | 1451 - 1506 |

Table 7.19. Conventional radiocarbon dates and calibrated ages on charcoal and marine shell from Area C (S8/16), Layer 2, Aotea.

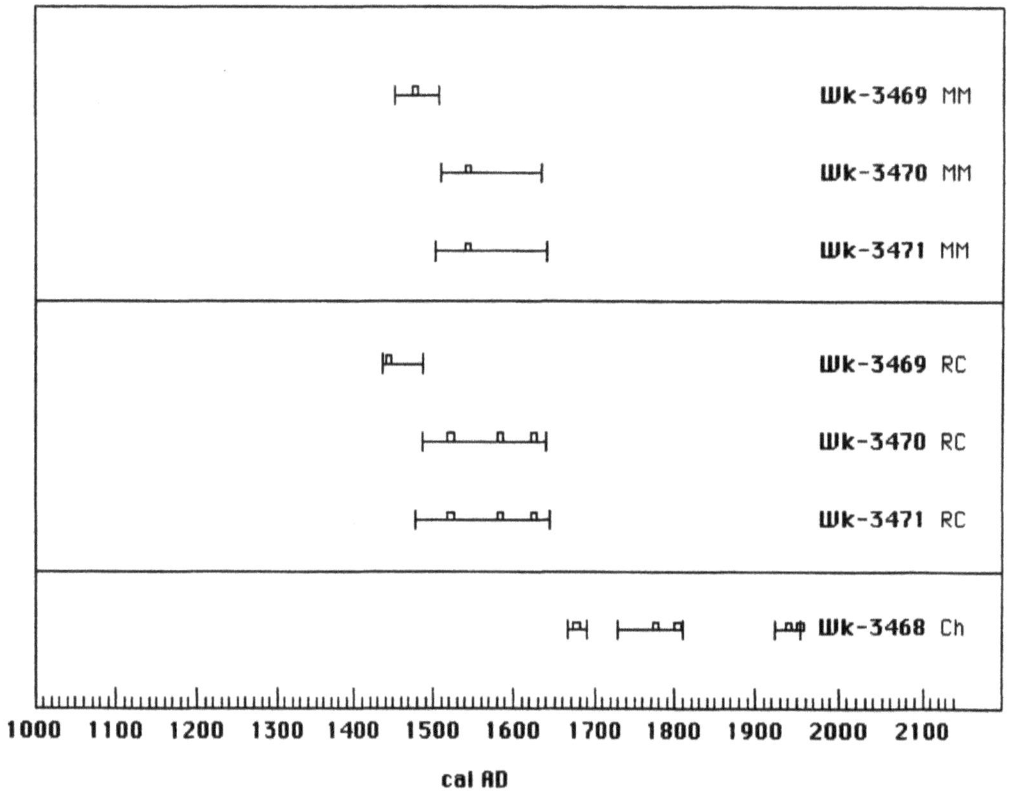


Figure 7.17. Calibrated radiocarbon ages from Area C (S8/16), Layer 2, Aotea. (MM = shell calibrated by Modelled Marine Curve; RC = shell calibrated using Reservoir Correction method; Ch = calibrated charcoal age).

## Post Depositional Disturbance: Historic and Prehistoric

### *Area A, Layer 2 (S8/171) and Area C, Layer 2 (S8/16), Aotea*

Radiocarbon anomalies from Areas A (S8/171) and C (S8/16) at Aotea are probably attributable to post-depositional disturbance (Tables 7.18 - 7.19; Figures 7.16 - 7.17). Area A was disturbed by ploughing and building activities in historic times (Grey 1994:5; McGhie 1994:29). The marine shell radiocarbon ages from Layer 2 may represent repeated occupations of the site in prehistory through to the modern era (Table 7.18; Figure 7.16). At Area C, the marine shell radiocarbon results are older than the charcoal determination from an identical context (Table 7.19; Figure 7.17). This is probably because of post-depositional reworking of shell midden from a previous occupation on the living terrace. It appears that the inhabitants of Area C levelled the terrace using midden from an earlier occupation, thus the dated samples may not date the Area C occupation accurately (Clough pers. comm.[26]).

## Mixing of Natural and Cultural Material

At two archaeological sites, there appears to have been mixing of natural marine shell deposits with cultural material. This caused the shell dates to be excessively old in comparison to the charcoal dates. The affected sites are Site 95 on Ponui Island, and Sawpit Point (N26/214), Awaroa.

[26] Dr. R. Clough, Department of Anthropology, University of Auckland.

| Lab No. | Sample Type | Layer | CRA | Cal Age AD 1σ (Atmospheric Curve) | Cal Age AD 1σ (Marine Curve) |
|---|---|---|---|---|---|
| Wk-3599 | Charcoal | B | MODERN~ | MODERN | |
| Wk-3600 | *Austrovenus stutchburyi* | B | 770 ± 40 | 1445 - 1522<br>1583 - 1625 | 1479 - 1545 |
| Wk-3601 | Charcoal | D | 600 ± 45 | 1329 - 1349<br>1391 - 1425 | |
| Wk-3602-1 | *Austrovenus stutchburyi* | D | 1560 ± 40 | 780 - 791<br>803 - 853<br>857 - 899<br>919 - 977 | 774 - 877 |
| Wk-3602-2 | *Austrovenus stutchburyi* | D | 1480 ± 50 | 895 - 924<br>940 - 999<br>1013 - 1016 | 862 - 975 |

Table 7.20. Conventional radiocarbon dates and calibrated ages on charcoal and marine shell from Site 95, Layers B and D, Ponui Island. (~though Wk-3599 is MODERN, the CRA of 187 ± 40 has been calibrated in figure 7.18 to illustrate the differences between calibrating charcoal CRA using the atmospheric calibration curve, and calibrating shell CRA using the modelled marine curve.).

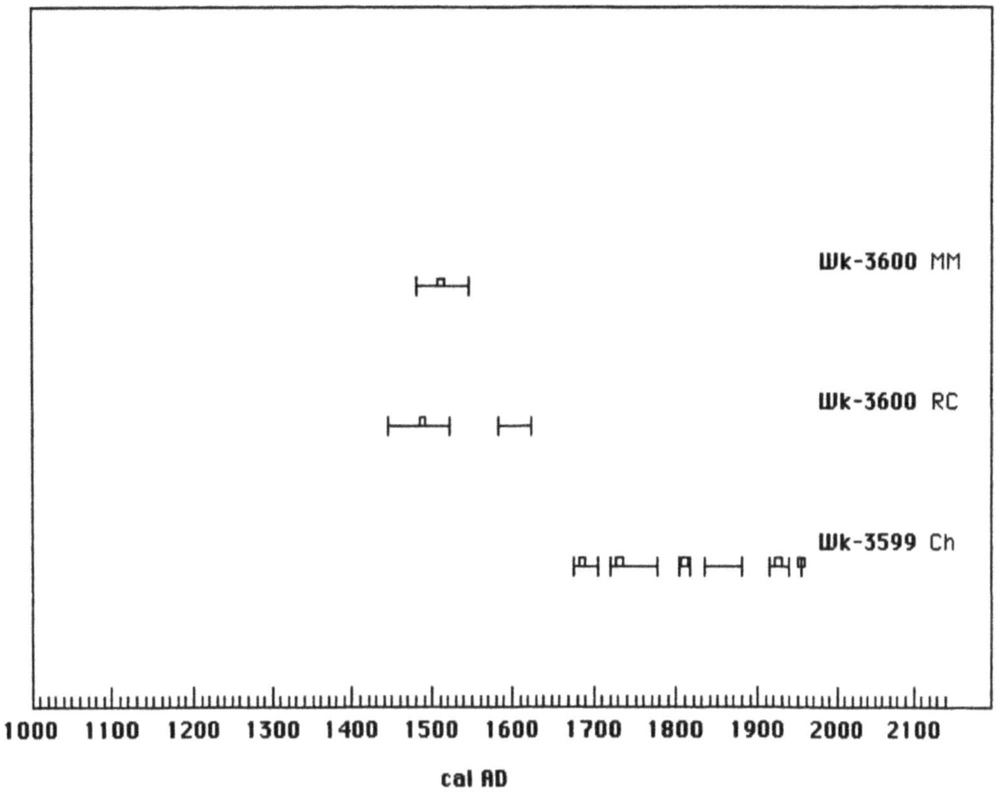


Figure 7.18. Calibrated radiocarbon ages from Site 95, Layer B, Ponui Island. (MM = shell calibrated by Modelled Marine Curve; RC = shell calibrated using Reservoir Correction method; Ch = calibrated charcoal age).

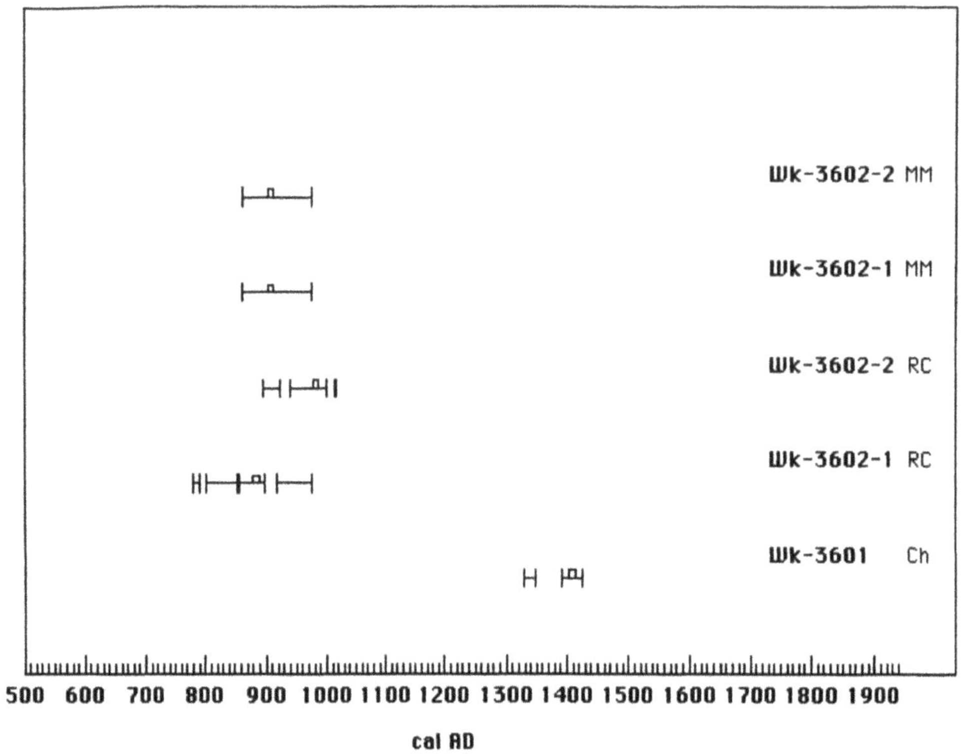


Figure 7.19. Calibrated radiocarbon ages from Site 95, Layer D, Ponui Island. (MM = shell calibrated by Modelled Marine Curve; RC = shell calibrated using Reservoir Correction method; Ch = calibrated charcoal age).

***Site 95, Layers B and D, Ponui Island***

The unusual nature of the midden matrix in both Layers B and D at Site 95 suggested to the excavators that some of the shell remains may have been deposited naturally (see Chapter 6). The presence of a single shellfish species in the layers set this cultural layer apart from others on the island, as all other cultural deposits contained a wide variety of species. Radiocarbon dating of Layers B and D at Site 95 confirmed that the marine shell deposit of *Austrovenus stutchburyi* (Wk-3600, Wk-3602-1, Wk-3602-2) was much older than other nearby sites, and therefore probably not cultural (Table 7.20; Figures 7.18 - 7.19). Both cultural layers show a marked discrepancy between calibrated shell and charcoal dates. Though the Layer B calibrated shell age does fall within the accepted time for human presence on Ponui, the discrepancy between this age and the charcoal date is very large. The depth of Layer B below ground surface (see Chapter 6), implies occupation within the last 250 years, as indicated by other sites on Ponui such as Layer B at site N43/333. Also, the similarity of matrices of Layer B and Layer D, and the similar charcoal/marine shell age discrepancies, imply that a similar process has affected the chronology of both sites. It is likely that natural beach shell was mixed with charcoal and hangi stones of cultural origin from subsequent human occupation in both layers, as originally believed by the excavators (see Chapter 6). This experience must serve as a cautionary note for archaeologists choosing samples for radiocarbon dating. The palaeoenvironment of Site 95 was similar to the rest of Ponui Island. At least four shellfish species could have been exploited by the prehistoric occupants, judging by the spectrum of shell midden refuse at nearby sites. Finding a midden containing only one shellfish species was therefore unusual. The location of the site directly on a beach front and still within the high tide mark could, in hindsight, be seen as a clue that the deposits may be prone to the incorporation of natural shell. Neither hard water effects or upwelling are variables which could serve to affect shell ages in this region.

I compared a sample of *Austrovenus stutchburyi* shells from Site 95, with *Austrovenus stutchburyi* from other prehistoric sites dated on the island, but could ascertain no significant taphonomic differences. Flessa *et al.* (1993:281,284;1994) have noted that using taphonomic indicators to distinguish between recent and ancient marine shells is very difficult, because preservation depends greatly on the depositional environment. Because of the sheltered bays of the western shore of Ponui Island from which archaeological sites were sampled, distinguishing between old beach shell and recently dead shellfish would be difficult. Flessa *et al.* (1993:384) also found radiocarbon ages ranging from modern to 3569 years BP on marine shells sampled from the surface of a beach flat, suggesting that the incorporation of natural shells in archaeological sites can induce significant errors in radiocarbon dating, by hundreds or thousands of years.

At Site 95, the date of occupation of Layers B and D are therefore probably represented most accurately by the charcoal radiocarbon ages.

***Sawpit Point (N26/214), Awaroa***

Sub-fossil beach shell also affected the accuracy of radiocarbon dating of Layer L1C at Sawpit Point (Wk-4032) (Table 7.21; Figure 7.20). Barber (1994) believed that some of the *Austrovenus stutchburyi* from this layer was sub-fossil (see Chapter 6), but determining which *Austrovenus stutchburyi* was sub-fossil, and which were cultural on taphonomic grounds was difficult. Some of the *Austrovenus stutchburyi* shells were rounded giving an appearance of age, but others were only partially rounded and fresh in appearance. Because of the coarse nature of the granite sands in this location, wear and rounding of marine shells would not necessarily indicate significant time lapsed since death. The radiocarbon date on *Austrovenus stutchburyi* (Wk-4032) confirmed Barber's (1994) assertion that this shellfish species in the site was probably sub-fossil.

| Lab No. | Sample Type | Layer | CRA | Cal Age AD 1σ (Atmospheric Curve) | Cal Age AD 1σ (Marine Curve) |
|---|---|---|---|---|---|
| Wk-4028 | Charcoal | L1B | 350 ± 40 | 1519 - 1594<br>1622 - 1650 | |
| Wk-4033 | *Paphies australis* | L1B | 830 ± 40 | 1432 - 1454<br>1458 - 1479 | 1446 - 1498 |
| Wk-4030 | *Austrovenus stutchburyi* | L1B | 790 ± 40 | 1441 - 1495<br>1601 - 1616 | 1467 - 1528 |
| NZ-7701i | Charcoal | L1C | 310 ± 110 | 1447 - 1671<br>1781 - 1799<br>1945 - 1953 | |
| Wk-4029 | Charcoal | L1C | 410 ± 40 | 1450 - 1528<br>1558 - 1631 | |
| Wk-3099 | *Paphies australis* | L1C | 800 ± 50 | 1437 - 1495<br>1601 - 1616 | 1457 - 1527 |
| Wk-4032 | *Austrovenus stutchburyi* | L1C | 1870 ± 50 | 541 - 642 | 453 - 593 |
| Wk-4031 | *Lunella smaragda* | L1C | 800 ± 40 | 1439 - 1492 | 1462 - 1520 |

Table 7.21. Conventional radiocarbon dates and calibrated ages on charcoal and marine shell from site N26/214, Layers L1B and L1C, Sawpit Point, Awaroa.

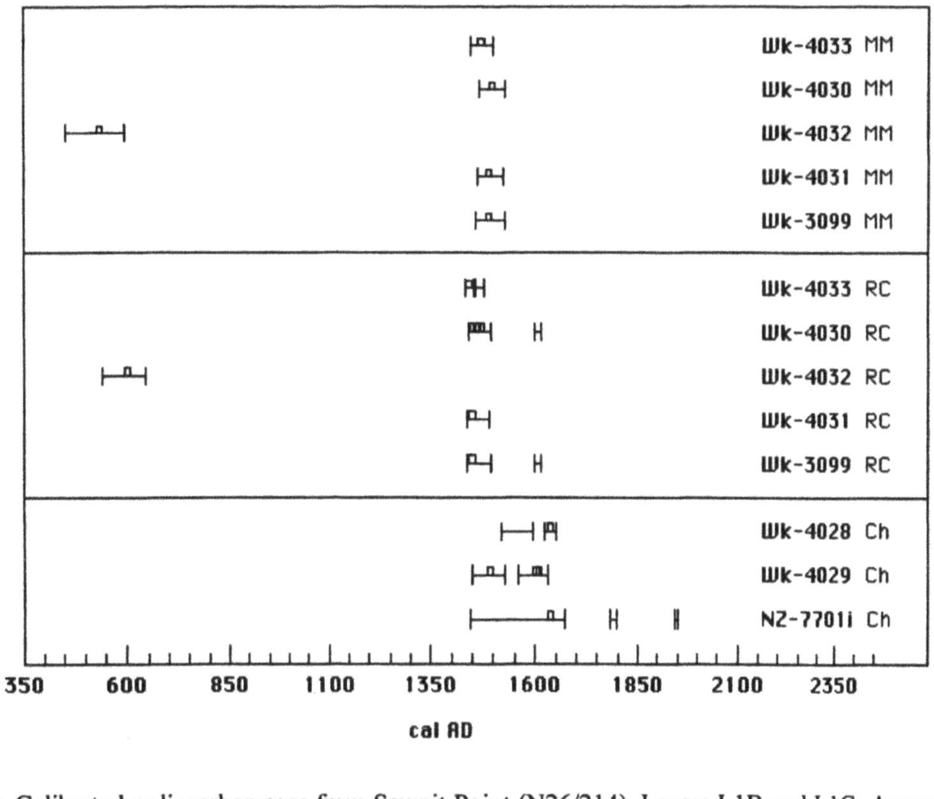


Figure 7.20. Calibrated radiocarbon ages from Sawpit Point (N26/214), Layers L1B and L1C, Awaroa, showing influence on chronology by incorporation of sub-fossil shell (Wk-4032). (MM = shell calibrated by Modelled Marine Curve; RC = shell calibrated using Reservoir Correction method; Ch = calibrated charcoal age).

### Summary

Each of the archaeological sites experiencing age anomalies discussed above, illustrate the risks involved in using marine shell and charcoal for radiocarbon dating prehistoric contexts. They also emphasis the importance of considering factors outside the radiocarbon method, eg. post-depositional disturbance. These sites may be excluded from further analysis on the basis of the reasons set out above, but the factors which caused their exclusion should be considered by archaeologists when endeavouring to date prehistoric deposits.

## DISCUSSION AND CONCLUSIONS

### Radiocarbon Dating of Marine Shell and Charcoal

The consistency in radiocarbon dates between different shellfish species is immediately apparent after a perusal of the shell ages from the archaeological sites presented above (Tables 7.1 - 7.16; Figures 7.1 - 7.14). This suggests that the shellfish species *Paphies australis, Austrovenus stutchburyi, Paphies subtriangulatum, Lunella smaragda, Venerupis largillierti, Perna canaliculus, Crassostrea glomerata, Cominella adspersa* and *Cominella virgata* are all reliable at this level of precision for radiocarbon dating New Zealand prehistoric sites. It also appears that the habitats and life-ways of these species have not effected accuracy. Those shellfish species from open beach environments provided similar radiocarbon ages to species from both rocky shore and estuarine environments (see Chapter 4).

In addition, the calcareous content of rocks surrounding the archaeological sites does not appear to have affected radiocarbon dates of shellfish. Even at Kokohuia, where a hard water effect was considered a source for error, none was detected.

Multiple charcoal radiocarbon dates from a number of sites used in this study were also consistent, presumably because the samples were rigorously identified to twigs of short-lived species prior to dating. In some instances, however, charcoal ages from the same deposit, did vary markedly in age between the youngest and oldest radiocarbon date (see Kokohuia). Such variations on charcoal samples from the same provenance can mean the difference between a single calibrated charcoal radiocarbon date agreeing or disagreeing, with a paired calibrated marine shell date.

The variation in multiple charcoal determinations could either be due to statistical variation or inbuilt age. It appears from this analysis of charcoal and marine shell radiocarbon dates, that the variations in charcoal ages are probably due to non-systematic inbuilt age which may occur even in short-lived species. Anderson (1991:780-781) has noted that charcoal from twigs of short-lived species may still retain an age of up to 50 to 150 years or greater, and both Anderson (1991) and Simpson (pers. comm.) have suggested that the speed of decay is related to the local environment and to species (see Chapter 4). It is possible, therefore, that past environmental conditions at Kokohuia and Ponui Island could have brought about slow decay of dead wood, meaning that some of the wood used by the inhabitants at Kokohuia and site N43/333, Layer B on Ponui Island, could have been present in the environment for a long period of time before use. Alternatively, charcoal identified as twigs of short-lived species could have been mis-identified, and the samples may have contained either long-lived wood, or old wood.

Selecting charcoal from tree and shrub species known to have been used only when green by prehistoric Maori may be a solution to further reducing inbuilt age of charcoal samples. Examples of these tree and shrub species are:

- Manuka, Kanuka and Kowhai used as firewood or for making tools (Clifton 1990:77-78);
- Tawa as fuel, as well as small green branches from other species burnt to keep insects away (Taylor 1994:13);
- Manatu and Ribbonwood bark used for making seine net ropes (Taylor 1994:13);
- Puriri seeds for dyes (Taylor 1994:13);
- Miro, Hinua and Tawa exploited for oil and as food sources (Taylor 1994:13);
- Tree and shrub species used as medicines (see Riley 1994).

The presence of these species in a site would infer immediate use by the prehistoric inhabitants, thus inbuilt age may be significantly reduced. This would also mean that archaeologists may be able to accept twigs from medium to long-lived tree species more often for dating.

The analysis of paired marine shell and charcoal radiocarbon ages, suggests that marine shell will yield reliable and consistent radiocarbon determinations. Certainly, they show greater reproducibility than charcoal. Whereas multiple charcoal radiocarbon dates from the same stratigraphic context can show variation, multiple marine shell ages on a variety of shellfish species appear to provide a much tighter chronology. This is important because it suggests that a single marine shell radiocarbon age will reflect the time at which a prehistoric event took place more accurately than a single charcoal radiocarbon age. The only risk identified in this thesis in using marine shell to radiocarbon date archaeological sites, is the incorporation of shell from natural beach deposits. In this study, however, both sites which were effected by natural beach shell, Site 95 on Ponui Island and Layer L1C at Sawpit Point, were identified as problematic prior to dating.

### Comparison of Marine Shell Calibration Methods

This analysis found that marine shell CRAs calibrated using the modelled marine calibration curve, more often agreed with the paired calibrated charcoal age, than shell dates calibrated using the reservoir correction technique. The reservoir corrected calibrations appeared in some cases too old by approximately 20 years, compared to the modelled marine curve calibrations, and calibrated charcoal dates. Additionally, the age ranges associated with the modelled marine curve calibrations, more often lay within the age ranges of the paired calibrated charcoal dates.

Whether the calibrated age ranges produced by the modelled marine curve reflect the chronology of the archaeological sites concerned accurately, may only be evaluated by considering the following questions:

1) Does the modelled marine curve successfully simulate changes in R(t) through time?
2) Do the 'tight' calibrated age ranges produced by the modelled marine curve provide a true reflection of the date of the prehistoric event in question, based on the calibrated charcoal dates, reservoir corrected calibrated shell dates, and the archaeological evidence?

### *Calculating R(t) and ΔR*

In considering these questions, the reservoir ages R(t) and ΔR values for each of the radiocarbon dated archaeological sites above, have been calculated so as to provide additional data to that already presented (see Chapter 5) (Table 7.22).

For five archaeological deposits, multiple possible ΔR values resulted during calculation. This is because the Stuiver and Braziunas (1993) conversion curve used to derive a modelled marine radiocarbon age from a charcoal radiocarbon age for ΔR calculation, incorporates oscillations in atmospheric $^{14}C$ through time (see Chapter 5). This causes 'wiggles' to occur in the curve (see Figure 5.3). Hence, sometimes during the conversion of a single charcoal radiocarbon date to a modelled marine age, 'wiggles' in the conversion curve are encountered, producing multiple possible modelled marine ages. For example, the single charcoal radiocarbon date from Area B, Aotea, of 220 ± 40 years BP, produces three possible modelled marine ages: 650 ± 40, 560 ± 40 and 470 ± 40 years BP (see Figure 5.3). This equates to three possible ΔR values of 7 ± 47, 90 ± 47 and 187 ± 47 years. Because it is difficult to determine which ΔR value is appropriate for these sites, ΔR have been calculated only for sites which do not experience multiple ΔR values (Table 7.22).

### *Modelling Changes in R(t) Through Time*

Pooled mean marine shell CRA versus pooled mean charcoal CRA for each site in table 7.22, have been plotted in figure 7.21. These plots illustrate change in R(t) through time (calculated for each site in table 7.22) (see Chapter 5). Also plotted in figure 7.21, is the reservoir correction line for New Zealand (*R* = 385) (see Chapter 5). When using the reservoir correction calibration method, marine shell CRAs are applied to this line to convert a shell date into an equivalent terrestrial age, before calibration on the atmospheric calibration curve. The modelled marine calibration curve for the period 200 to 700 years BP is also shown in figure 7.21 (Stuiver and Braziunas 1993:183). It is apparent from figure 7.21, that R(t) can vary markedly through time (Note: plots marked * have only one charcoal age available for R(t) calculation). However, although the modelled marine curve does appear to successfully simulate the temporal variability of R(t), the fixed reservoir correction can also be seen to account for this change. This observation is confirmed when a polynomial curve of order 3 (R^2 = 0.649) and a straight line linear regression (R^2 = 0.642) are fitted to the reservoir age data (Figure 7.22). When these graphs are compared with figure 7.21, they illustrate that neither the fixed reservoir correction or modelled marine curve more successfully accounts for change in R(t) through time based on this analysis.

One reason for the difficulty in ascertaining which curve more successfully simulates change in R(t) through time, is the small number of charcoal CRA from some sites. Single charcoal ages make it difficult to precisely calculate R(t) as one age may not be representative of the range of dates from a site. The differences between each marine shell calibration method are only apparent when marine shell ages are calibrated using both methods (as seen in the sites above),

| Site | Pooled Mean Charcoal CRA *(A(p))** | Pooled Marine Shell CRA *(A(p))** | R(t) | ΔR value |
|---|---|---|---|---|
| Kokohuia Layer 2 | 370 ± 18 | 744 ± 18 | 374 ± 25 | - |
| Kokohuia Layer 3 | 373 ± 23 | 734 ± 18 | 361 ± 29 | - |
| Kokohuia Layer 4 | 435 ± 20 | 770 ± 20 | 335 ± 28 | -80 ± 28 |
| Kokohuia Layer 5 | 450 ± 17 | 815 ± 19 | 365 ± 25 | -45 ± 25 |
| Ligar Bay N26/95 | 455 ± 28 | 856 ± 14 | 401 ± 31 | -4 ± 31 |
| Ligar Bay N25/26 | 390 ± 40 | 742 ± 18 | 352 ± 44 | - |
| Ponui Is. N43/1 | 535 ± 48 | 843 ± 20 | 308 ± 52 | -77 ± 52 |
| Ponui Is. N43/333 | 420 ± 28 | 910 ± 28 | 490 ± 40 | 60 ± 40 |
| Ponui Is. Site 14 | 620 ± 40 | 917 ± 21 | 297 ± 45 | -63 ± 45 |
| Ponui Is. Site 27 | 470 ± 40 | 780 ± 28 | 310 ± 49 | -110 ± 49 |
| Sawpit Point (N26/214) | 376 ± 27 | 805 ± 21 | 429 ± 34 | - |
| Area B (S8/171), Aotea | 220 ± 40 | 657 ± 23 | 437 ± 46 | - |
| Pooled Mean *(A(p))* | | | 367 ± 10# | -42± 25† |

Table 7.22. Reservoir ages and ΔR values for archaeological sites dated in this study with acceptable pairs of charcoal and marine shell radiocarbon dates. Each site has had its multiple charcoal and marine shell conventional radiocarbon ages pooled (*A(p)*) before reservoir correction and ΔR values have been calculated. All pooled ages are statistically indistinguishable. * = Ward and Wilson (1978). # = Test statistic: $T = 23.01$; $\chi^2_{11:0.05} = 19.68$. † = Test statistic: $T = 12.46$; $\chi^2_{6:0.05} = 12.59$ (Largest of weighted or unweighted 'scatter' mean σ given for final pooled mean ΔR value rounded up to 5 years as recommended by Stuiver *et al.* (1986:982;1993:155)). Sites not showing ΔR values are where multiple ΔR are possible (see above).

where the influence of the calibration curves on the shell dates can be clearly seen (see below).

Some observations may be made, however, with regard to this data. It can be seen that the R(t) values calculated for the archaeological sites sampled in this study are statistically distinguishable, possibly indicating that a fixed reservoir correction is not applicable for all of the New Zealand prehistoric sequence (Table 7.22). Also, application of the modelled marine curve calibration method for New Zealand is dependent on the local ΔR value being non time-dependant. The ΔR value of -42 ± 25 calculated for the archaeological sites presented in table 7.22, is statistically indistinguishable from the current value of -25 ± 15 ($A(p)$ = -30 ± 13 ($T$ = 0.73; $\chi^2_{1:0.05}$ = 3.84)) (Higham and Hogg 1995). This indicates that there has been no temporal variation in ΔR over the past 700 years for New Zealand, and that use of the Higham and Hogg (1995) ΔR value in association with the modelled marine curve, should successfully calibrate marine shell CRAs. From the sites presented above, where there is agreement between the modelled marine curve calibrated shell dates, and paired calibrated charcoal dates, this has shown to be the case.

***Simulating Marine Shell Calibration and Differences between Calibrated Age Ranges***

McFadgen *et al.* (1994) have illustrated that 'wiggles' in the atmospheric calibration curve can make it difficult to date New Zealand prehistory using charcoal, especially after *ca.* 1500 cal AD. This causes an artificial spreading of calibrated data, and thus multiple age ranges. Shell ages calibrated using the reservoir correction method also incorporate this phenomena, as illustrated in the sites analysed above. The modelled marine curve calibration method has been shown to avoid multiple age ranges because of the smoothness of the Stuiver and Braziunas (1993) curve. The analysis of shell ages calibrated using both marine shell calibration methods above, indicates that the calibration methods only begin to differ significantly after 1500 cal AD. This affects marine shell CRAs less than 800 ± 40 years BP.

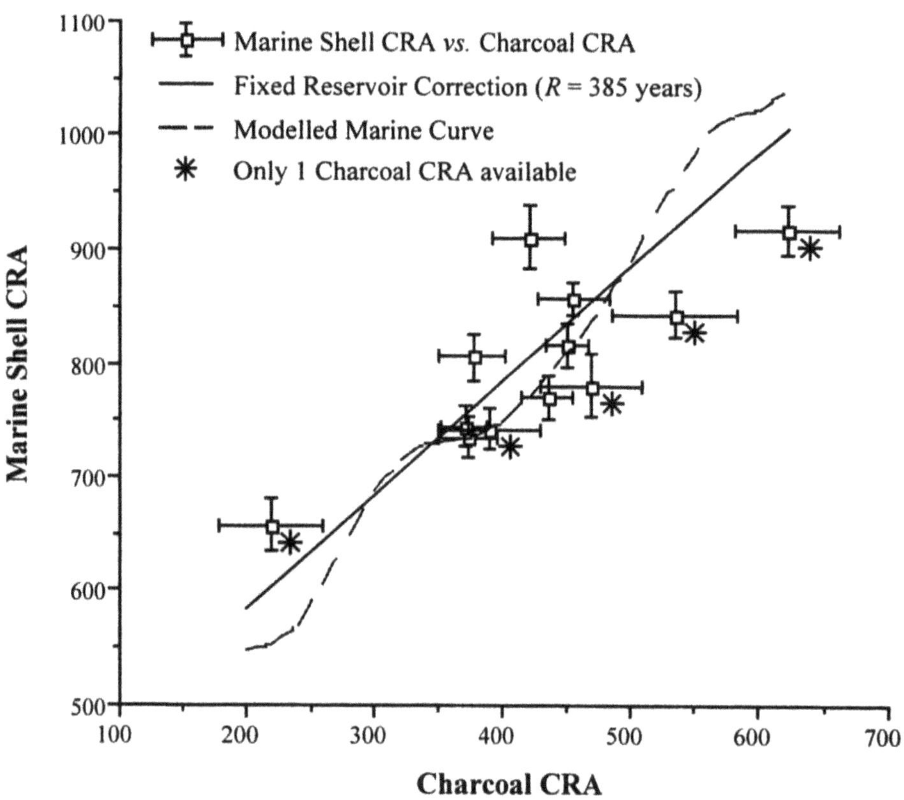


Figure 7.21. Graph showing pooled mean marine shell CRA vs. pooled mean charcoal CRA calculated from the archaeological sites in Table 7.22, plotted against the fixed reservoir correction line for New Zealand, and the Stuiver and Braziunas (1993) modelled marine curve for the period 200 to years 700 BP.

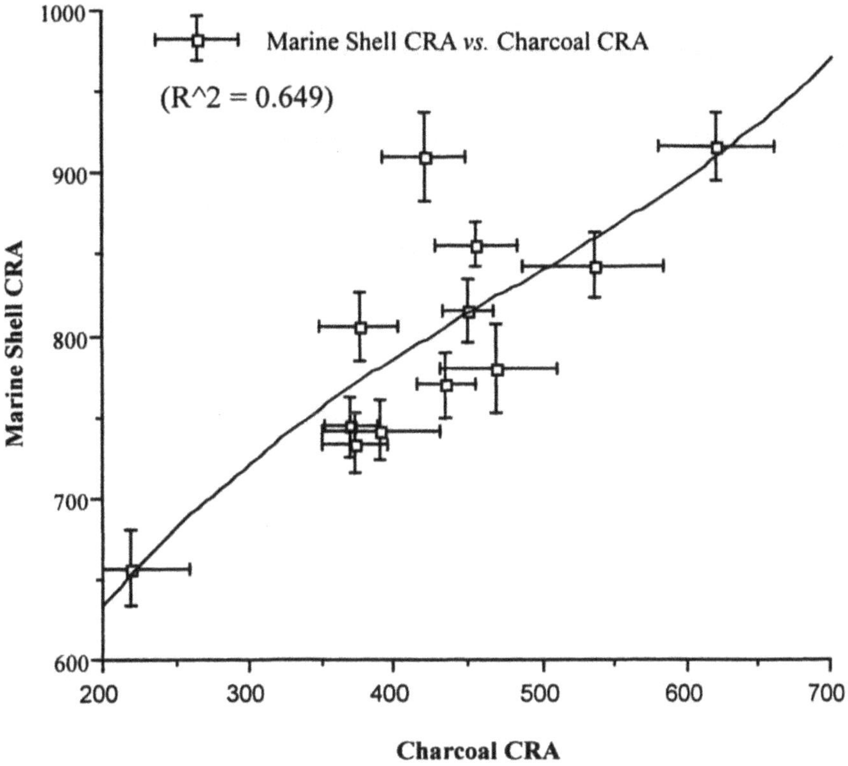


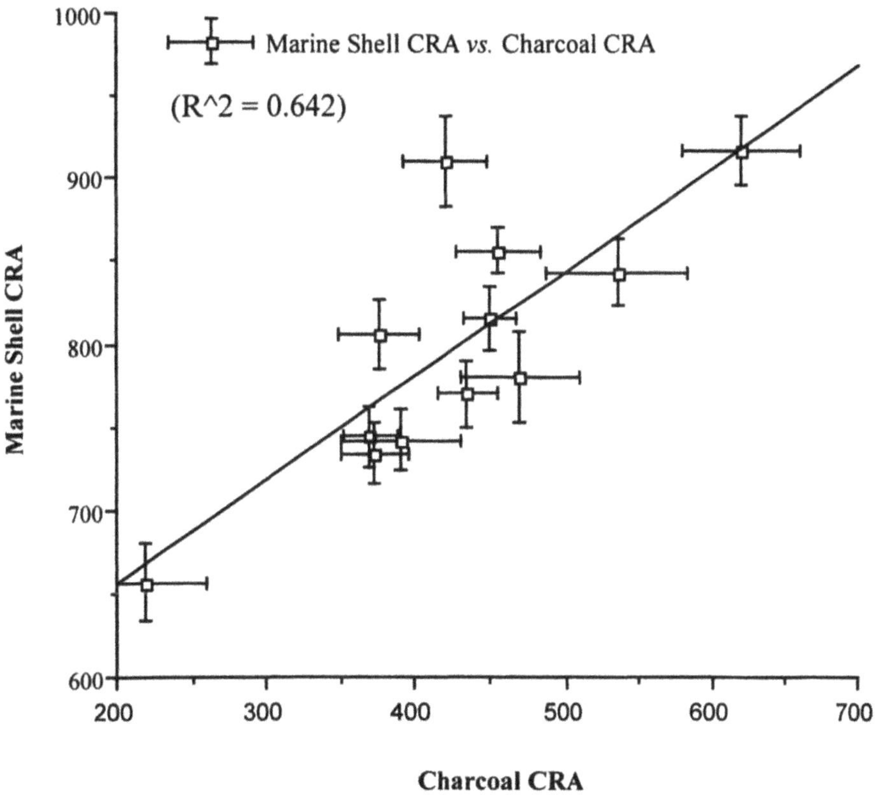


Figure 7.22. Polynomial (order 3) curve (top) and straight line linear regression (bottom) fitted to R(t) values calculated for the archaeological sites dated in this study.

When conventional marine shell radiocarbon dates older than 800 ± 40 years BP are calibrated using either shell calibration method, the age ranges are in close agreement. The calibrated marine shell radiocarbon dates from Kokohuia Layer 5 (Figure 7.4); Ponui Island sites N43/1, Layer D (Figure 7.7), N43/333, Layer D (Figure 7.8), and 14 (Figure 7.9); site N25/95, Ligar Bay (Figures 7.11 and 7.12); and Sawpit Point (N26/214) (Figure 7.14) illustrate this.

Conventional marine shell radiocarbon ages younger than 800 ± 40 BP, however, fall into the calibrated period of *ca.* 1500 cal AD onwards, where two significant 'wiggles' in the atmospheric calibration curve occur (Figure 7.23). This results in a spread in some reservoir corrected calibrated shell age ranges. This is particularly noticeable at Area B (S8/171), Aotea (Figure 7.6); Kokohuia Layers 2 to 4 (Figures 7.1 - 7.3); Site 27, Ponui Island (Figure 7.10); and site N25/26, Ligar Bay (Figure 7.13).

The effect of these 'wiggles' when calibrating shell ages younger than 800 ± 40 years BP using the reservoir correction method, can make the interpretation of archaeological events in the post-1500 AD period of New Zealand prehistory difficult (see McFadgen *et al.* 1994). This can be illustrated by simulating calibrated data for shell determinations using both calibration methods (Figures 7.24 and 7.25). The simulation begins with a conventional marine shell radiocarbon age of 800 ± 40 years BP (reading from the bottom of the page, Figure 7.25). This age is first calibrated on the atmospheric curve after subtraction of the New Zealand reservoir correction of 385 years (giving a reservoir corrected age of 415 ± 40 BP). Second, the 800 ± 40 years BP shell age is calibrated using the modelled marine calibration curve (Figure 7.25). The simulation ends with the calibration of a conventional marine shell radiocarbon age of 610 ± 40 years BP (Figure 7.24).

For the marine shell conventional radiocarbon age of 800 ± 40 years BP, the reservoir corrected calibrated age is slightly older than the modelled marine curve calibrated $^{14}C$ date, but both have 'tight' age ranges (Figure 7.25). From 790 ± 40 years BP to 760 ± 40 years BP, however, the age ranges produced by the reservoir correction calibration method begin to widen because of a 'wiggle' in the atmospheric calibration curve (Figure 7.25). At 750 ± 40 years BP,

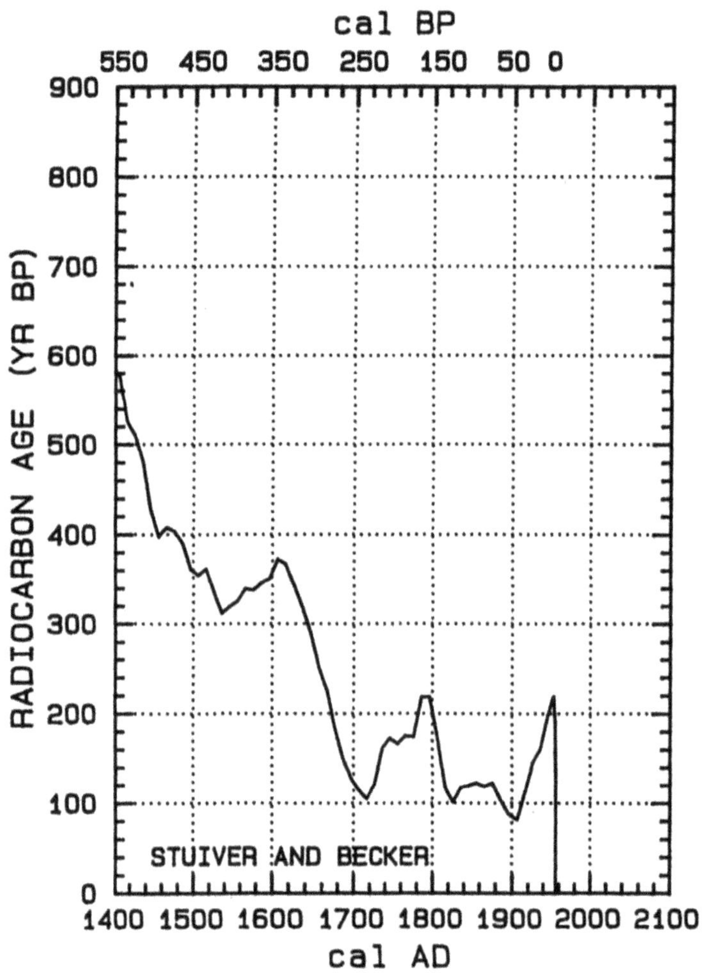


Figure 7.23. Stuiver and Becker (1993) decadal atmospheric calibration curve for the period 1400 to 1950 cal AD. 'Wiggles' in this curve increase after 1500 cal AD, significantly affecting the calibration of marine shell CRAs <800 ± 40 BP when using the reservoir correction calibration method.

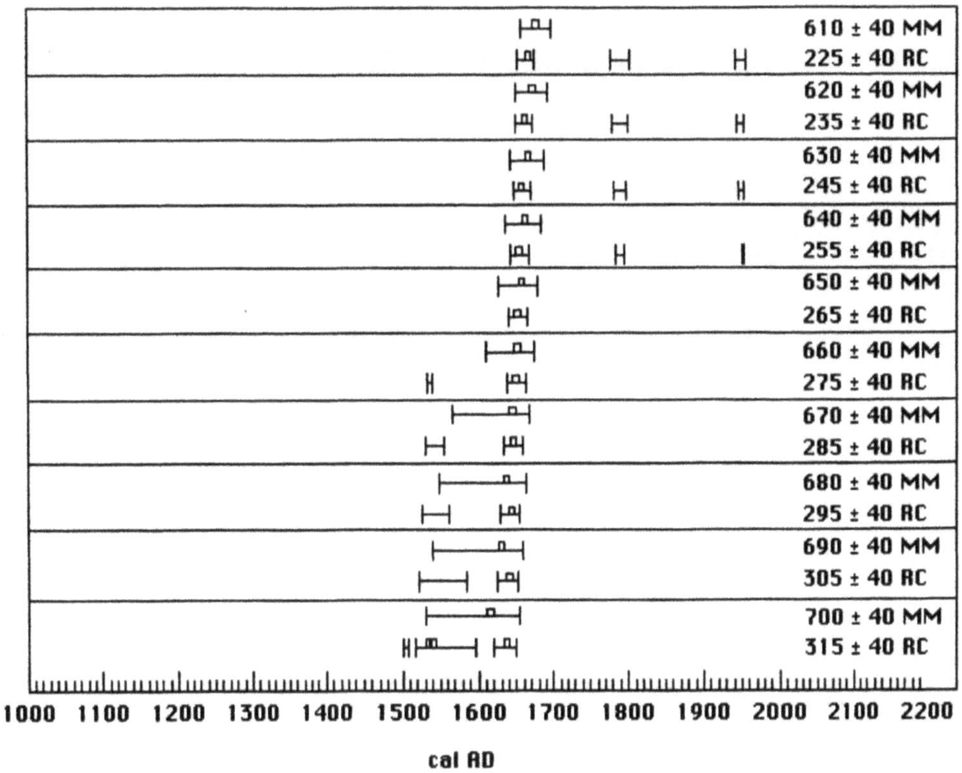


Figure 7.24. Simulated radiocarbon age calibrations using both marine shell calibration methods beginning with a marine shell conventional radiocarbon age of 700 ± 40 years BP. (MM = shell CRA calibrated using modelled marine curve; RC = shell CRA calibrated using reservoir correction method.).

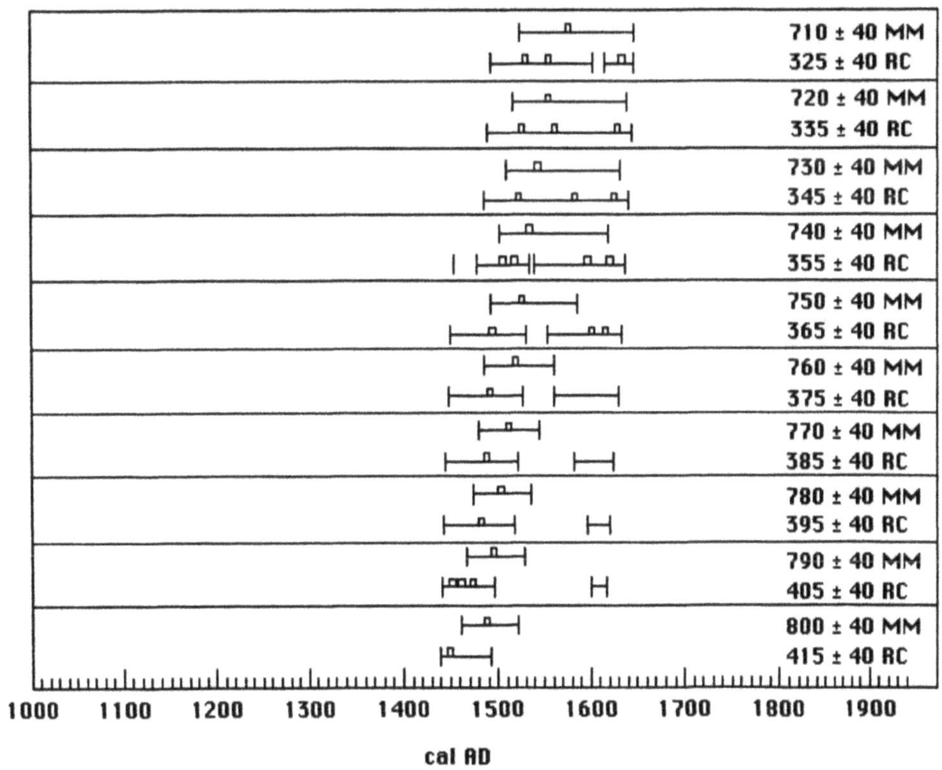


Figure 7.25. Simulated radiocarbon age calibrations using both marine shell calibration methods beginning with a marine shell conventional radiocarbon age of 800 ± 40 years BP. (MM = shell CRA calibrated using modelled marine curve; RC = shell CRA calibrated using reservoir correction method.).

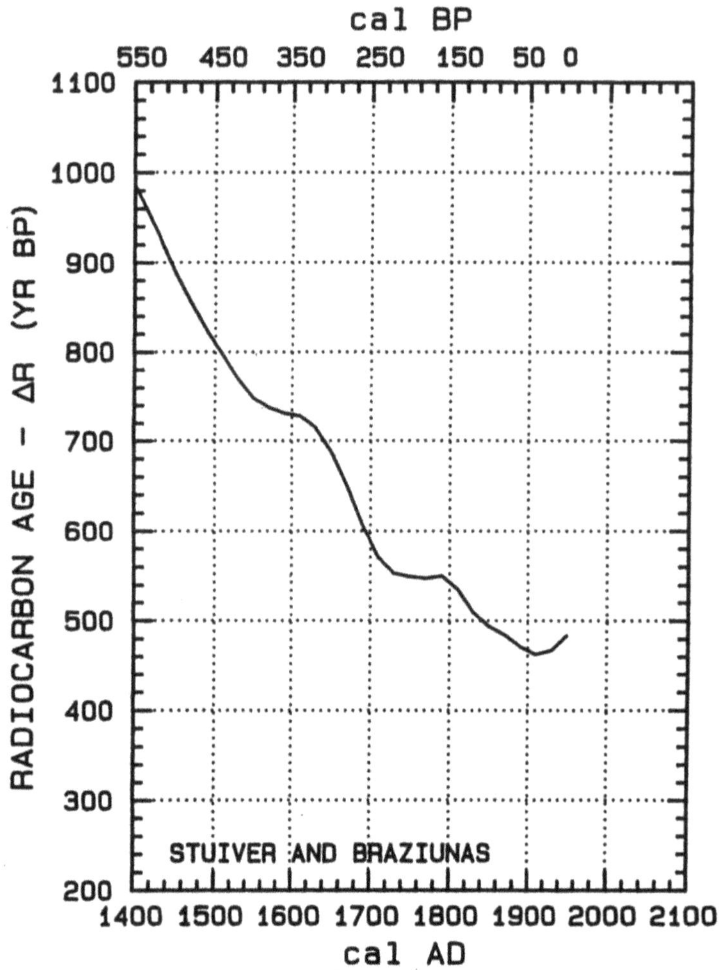


Figure 7.26. Stuiver and Braziunas (1993) modelled marine calibration curve for the period 1400 to 1950 cal AD.

marine shell CRAs calibrated using the modelled marine curve also begin to experience a widening of their calibrated age ranges in response to a flattening of this curve from an increase in $^{14}C$ levels for the mixed layer ocean (Figure 7.26). Between 740 ± 40 years BP and 670 ± 40 years BP, both calibration methods again produce similar age ranges, and at 650 ± 40 years BP they are in close agreement (Figures 7.24 and 7.25). After 650 ± 40 years BP, however, the atmospheric calibration curve once more experiences 'wiggles', causing age ranges produced by the reservoir correction technique to extend through to the modern period.

The reservoir correction and modelled marine curve calibration methods, therefore, only disagree for the calibration of marine shell dates between 790 ± 40 and 760 ± 40 years BP, and younger than 650 ± 40 years BP. For these shell CRAs, the modelled marine curve provides tighter chronological control, meaning that archaeological events younger than 1500 AD and beyond the time of Cook's arrival in 1769, may be more precisely dated.

The 'tight' modelled marine calibrations for this period do appear to be accurate in reflecting the age of archaeological events, when archaeological evidence is used as a relative age check on the site chronology. The calibrations from Area B, Aotea (Figure 7.6), and site N43/333, Layer B, Ponui Island (Figure 7.15) are examples where the archaeological evidence and excavator's interpretation of the time of occupation of the sites are more in line with the modelled marine curve calibrations, than those undertaken using the reservoir correction method. The shell CRAs of less than 650 ± 40 years BP from these sites calibrated using the reservoir correction method, imply that these sites could have been occupied up until the present day. The archaeologists who excavated these sites, however, believed that they were occupied before European arrival, but after 1600 AD. The modelled marine curve calibrations, therefore, appear to provide a more accurate indication of when occupation actually occurred.

Further testing of the calibration of marine shell dates from archaeological sites believed to have been occupied between 1500 AD and European arrival would further confirm use of the modelled marine curve for this period of New Zealand prehistory.

Do the calibrated age ranges obtained using the modelled marine curve calibration method accurately reflect the time at which an archaeological event took place? The analysis presented above suggests that the age ranges produced by the modelled marine curve are a true representation of when the archaeological events occurred in New Zealand prehistory.

This can be seen in the close agreement between modelled marine curve calibrated dates, and calibrated charcoal ages from identical stratigraphic horizons. In addition, the modelled marine curve appears to successfully simulate changes in R(t) through time (Figure 7.21). The $\Delta R$ value calculated from the sites analysed in this study is statistically indistinguishable from the current New Zealand $\Delta R$ value (Table 7.22), supporting the conclusion of Higham and Hogg (1995) and McFadgen and Manning (1990) of a singular lack of any temporal variation in this value over the last millennium in New Zealand waters. Also, the simulations show the closeness of both marine shell calibration methods, and that noticeable differences only occur with certain marine shell CRA in the post-1500 cal AD period, though these differences can be reduced by incorporating archaeological evidence.

# CHAPTER 8

# DISCUSSION AND CONCLUSIONS

## OVERVIEW

The reliability of a particular sample type for radiocarbon dating is an important variable for the prehistorian seeking to build accurate chronologies. A useful sample type should by definition be independently accurate and provide adequate reproducibility amongst a group of repeat measurements on the same material. Careful studies have shown that if these variables are not critically evaluated prior to radiocarbon dating analysis, significant errors may eventuate.

In Chapter 3, diverse opinions from archaeologists were presented concerning the reliability of different sample types for radiocarbon dating New Zealand archaeological sites. It was found that of the many opinions expressed, only a handful were based upon careful scientific analyses. Prior to 1976, the single most critical error concerned with dating was inbuilt age of charcoal, but it was rarely considered when comparisons with marine shell or bone chronologies were made. Additionally, comparisons between radiocarbon dates of different sample types from archaeological sites, were often undertaken without due consideration of the precise stratigraphic location of the samples involved, and without rigorous assessment of the reliability of the sample in dating a specific prehistoric event. In some cases, the post-depositional history of the sites was overlooked. After 1976, the identification of charcoal samples to twigs of short-lived species prior to dating improved reliability markedly. However, comparative studies after this time were affected by a failure of researchers to differentiate between samples comprising identified charcoal, and those which did not.

Charcoal was the only material considered reliable by many New Zealand archaeologists after 1976. The usefulness of dating marine shell, which became a commonly used sample type by archaeologists after this time, was still relatively unknown. Because of this, Anderson (1991), in his critical analysis of the New Zealand radiocarbon record, focussed principally upon the radiocarbon dates of charcoal. It was not until 1993, that the applicability of marine shell for radiocarbon dating New Zealand prehistory was closely examined (see Higham 1993), by comparing paired radiocarbon dates from *identified* charcoal, with marine shell. Two of the three sites studied by Higham (1993), showed agreement between the two samples types, but at Pleasant River, anomalies between marine shell radiocarbon dates on various shellfish species, and their paired charcoal radiocarbon ages were found. Higham (1993) initially suggested that the differences in radiocarbon age between shellfish species may have been due to estuarine shellfish incorporating young carbon from terrestrial run-off. Further radiocarbon dating revealed that the most likely explanation was that the site had suffered post-depositional disturbance (Higham pers. comm.).

It was found in Chapter 3 that over the last 20 years, many more shell samples have been dated than charcoal from New Zealand archaeological sites (in the Waikato laboratory since 1980, for instance, the ratio is 3:1 in favour of shell (Higham pers. comm.). Further examination of shell for dating was therefore seen to be important. A solution to the sample type dilemma was in turn proposed. This involved a further comparison between paired charcoal and marine shell radiocarbon dates from the same archaeological provenance. In contrast to most previous studies, though, all the charcoal would be carefully identified prior to radiocarbon dating as Higham (1993) had done, and a greater variety of shellfish species tested. Charcoal and marine shell pairs from archaeological sites from varying locations and environments around New Zealand would be selected for analysis. In addition, the two methods available to archaeologists today for calibrating conventional marine shell radiocarbon dates would be scrutinised to ascertain the reliability of each for New Zealand prehistory.

In Chapter 4, factors which can affect the radiocarbon dating of charcoal and marine shell samples were examined. The rate of decay of wood after death was found to be an additional variable in determining inbuilt age, even in supposedly reliable short-lived species. In certain environments and under the right depositional conditions, twigs may survive for many years and thus incorporate serious errors in radiocarbon dates. Particular tree species considered to be long-lived and not suitable for dating can decay rapidly after death, and may not necessarily be unsuitable for dating. The life expectancy of various native tree and shrub species was also found to be controversial. Certain short-lived species can possibly live for a far greater period of time than has been suggested.

Upwelling, the hard water effect, and the incorporation of organic (metabolic) carbon, are factors which may affect radiocarbon dates of shell. Examples from different contexts illustrating the effects of these errors were shown. Only one example of a marine hard water effect was found, involving shellfish living upon limestone strata, and was identified as possibly being species specific (Dye 1994). The most common environment for hard water was coastlines associated with large fresh water inputs, and enclosed fresh water lakes. Only two organic carbon effects were noted. In the case of Little (1993), the conclusion that organic carbon had affected the age of radiocarbon dates was considered unlikely because the charcoal samples used in her study were not identified. Inbuilt age may be considered a more likely explanation for Little's (1993) observed variations in radiocarbon measurements. Higham's (1993) suggestion that variation in radiocarbon age related to modern organic carbon incorporation into shellfish at Pleasant River has been addressed above.

An analysis of the two currently available marine shell calibration methods was undertaken in Chapter 5. The original method of calibrating marine shell, using a fixed reservoir correction (R) or marine shell standard (MSS), has caused problems in New Zealand archaeology. In New Zealand, for example, six different reservoir values were suggested by various researchers between 1972 and 1984, for correcting marine shell radiocarbon ages prior to calibration. This caused a variety of interpretive problems for New Zealand archaeologists in establishing when a prehistoric event occurred. Because the recommended reservoir correction varied from between 330 and 430 years, the age of contemporary archaeological deposits dated using marine shell sometimes differed in age by up to 100 years. Discrepancies between calibrated radiocarbon dates on charcoal and marine shell from the same location were also noted when using this method of correction. Another problem identified with the use of a fixed reservoir correction, was that it did not take into account changes in R(t) through time.

A new marine shell calibration method known as the Modelled Marine Curve was examined. The basis of this method is a Global Carbon Model (GCM) developed by Stuiver *et al.* (1986), which was based upon Oeschger *et al.'s* (1975) box diffusion model (BDM) that simulated the movement of carbon through the biosphere and world's oceans. The GCM model simulates changes in mixed layer $^{14}C$ levels through time, providing a smooth marine calibration curve due to the slow diffusion of $^{14}C$ from the atmosphere and deep ocean, into the mixed layer ocean. This gradual fluctuation in mixed layer ocean $^{14}C$ contrasts with high variability in atmospheric $^{14}C$, which causes 'wiggles' in the atmospheric calibration curve. The modelled marine curve enables the calibration of conventional marine shell radiocarbon dates, which accounts for changes in R(t) in the mixed layer ocean through time. Associated with this curve is a non time-dependant regional reservoir correction called $\Delta R$, calculated in a similar way as R(t), which accounts for local reservoir activity that diverges from the simulated ocean reservoir.

The new modelled marine curve has not been extensively tested with archaeological examples. A review of current research on this calibration method showed that methodology used in its investigation were often hampered by a lack of identification of the charcoal samples, and cursory attention given to the lifeways and habitats of various shellfish species dated. These same errors in methodology were those also being made by researchers calculating country specific *R*, and so similar problems were being encountered with the calculation of $\Delta R$. This therefore meant that is was difficult to accurately assess the usefulness of the modelled marine curve for New Zealand prehistory.

The 11 excavated archaeological sites sampled by the author to obtain paired charcoal and marine shell radiocarbon dates for this study were described in Chapter 6. These sites were excavated in regions containing varying amounts of associated calcareous geological strata, to consider the hard water effect. A variety of shellfish species from estuarine, sandy and rocky shore locations were sampled to examine the influence of environmental variables on the radiocarbon dating of marine shell.

The analysis of charcoal/marine shell paired radiocarbon dates in Chapter 7, resulted in 14 of the 19 archaeological deposits yielding calibrated marine shell radiocarbon ages which agreed with their paired calibrated charcoal radiocarbon dates when either marine shell calibration method was used.

For sites where the calibrated charcoal and shell dates were in chronological agreement, no differences in radiocarbon dates between different shellfish species were found. All shellfish species produced the same age for the archaeological event being dated. Though the shellfish species came from estuarine, sandy and rocky shore environments, this did not affect accuracy. The hard water effect was not detected in gastropods or bivalves living directly upon limestone strata, nor was it identified for estuarine species associated with riverine inputs which flowed through highly calcareous strata. Based on these observations, it was concluded that at least nine shellfish species commonly found in New Zealand archaeological sites, can be considered reliable for radiocarbon dating (Table 8.1).

Few problems were observed regarding the charcoal radiocarbon ages, but the problem of inbuilt age appeared to still be present. At one archaeological site, two cultural layers had been dated with four charcoal radiocarbon determinations. Although as a set, each group of charcoal ages was statistically indistinguishable, separately, the youngest and oldest results were statistically distinguishable. At another archaeological site, the calibrated charcoal radiocarbon date differed in age from its paired calibrated shell result by *ca.* 150 years (see below). Three important variables in the dating of charcoal samples were identified in the analysis of the charcoal dates. First, whether the forest type surrounding the sites in prehistory was dominated by long-lived or short-lived flora. Second, the possibility that various tree and shrub species thought to be short-lived may live for a longer period of time. Third, that variations in decay rates of twigs from short-lived trees and shrubs around the country, may involve the differential preservation of this material in the archaeological record.

| **Estuarine** | **Sandy Shore** | **Rocky Shore** |
|---|---|---|
| *Austrovenus stutchburyi* | *Paphies subtriangulatum* | *Cominella adspersa* |
| *Paphies australis* | | *Cominella virgata* |
| *Venerupis largillierti* | | *Crassostrea glomerata* |
| | | *Lunella smaragda* |
| | | *Perna canaliculus* |

Table 8.1. Shellfish species found to be reliable in this study for radiocarbon dating New Zealand prehistory.

When calibrating shell-charcoal paired dates, the modelled marine curve generally produced calibrated age ranges which were in closer agreement with the calibrated charcoal dates, than the reservoir corrected calibrated shell dates. Shell dates calibrated using the reservoir correction method sometimes produced age ranges older than the modelled marine curve calibrated dates by approximately 20 years. Also, the reservoir corrected calibrated marine shell dates experienced multiple and expanded age ranges, compared to the 'tight' calibrations produced by the modelled marine curve. The difference between both marine shell calibration methods, was identified as being due to the 'wiggles' in the atmospheric calibration curve causing expanded age ranges, when calibrating shell dates using the reservoir correction method. The smoothness of the modelled marine calibration curve, avoids these 'wiggles', and therefore produces more constrained age ranges.

For the sites where calibrated marine shell and charcoal dates were in chronological disagreement, post-depositional disturbance, the mixing of natural beach shell deposits and cultural midden, and inbuilt age, were found to be responsible. At one site, ploughing and the building of a homestead mixed cultural material from two differing periods. In a second site, midden from a *pa* terrace was used to fill a pit and level a living terrace later in prehistory, causing a mix of older and contemporary cultural material. Incorporation of natural beach or sub-fossil marine shell with cultural material, caused large age variations between marine shell and charcoal dates from three archaeological deposits. However, the archaeological sites where this situation occurred were already identified by the excavators prior to dating. At one site, briefly mentioned above, the calibrated charcoal age was *ca.* 150 years older than the paired calibrated marine shell dates. This was proposed as being due to inbuilt age based on: 1) the age of the associated shell dates; 2) that a lower layer showed charcoal and marine shell chronological agreement, and these dates were only marginally younger than the disputed charcoal age, and 3) that the dubious charcoal age was too old based on stratigraphic grounds.

In Chapter 7, the accuracy of the modelled marine curve in producing realistic calibrations for marine shell dates was tested. To illustrate changes in R(t) through time, shell CRAs were plotted against their paired charcoal CRAs for sites where the shell/charcoal paired dates were in chronological agreement after calibration. Also plotted was the fixed reservoir correction line for New Zealand ($R$ = 385 years) and the modelled marine calibration curve for the period 200 to 700 BP. It was found that both calibration methods closely accounted for the temporal variability of R(t), and it was difficult to discern from the graphs which shell calibration method was the most applicable. This was because of the large variations in R(t) calculated from some sites due to only one or two charcoal radiocarbon ages being available. However, the R(t) values from the sites were found to be statistically distinguishable. Additionally, the pooled mean $\Delta R$ value calculated from the archaeological sites in this study, was indistinguishable from the current New Zealand $\Delta R$ value of -25 ± 15 (Higham and Hogg 1995). This in turn supported the view that $\Delta R$ was non-time dependent for the New Zealand prehistoric sequence, which is a central theme to the applicability of the modelled marine curve in the calibration of marine shell radiocarbon dates.

An investigation of the accuracy of the modelled marine curve in age calibration was also undertaken. In Chapter 7, it was shown that shell ages greater than 800 ± 40 years BP calibrated using both marine shell calibration methods showed close agreement. The only differences were that calibrated dates using the reservoir correction method were slightly older, and marginally wider in their age ranges. Conventional marine shell radiocarbon ages less than 800 ± 40 years BP calibrated using the reservoir correction technique, however, yielded large calibrated age ranges because of an increase in the 'wiggles' in the post-1500 cal AD period of the atmospheric calibration curve.

A calibration simulation was developed to investigate whether this phenomenon was dependent upon radiocarbon age. The simulation began at 800 ± 40 years BP and reduced in decadal intervals to 610 ± 40 years BP. This simulation showed that significant disagreements occurred between the calibration methods at only particular points in the post-1500 cal AD period. These disagreements occurred for shell CRA between 790 ± 40 and 750 ± 40 years BP, and <650 ± 40 years BP. For these shell ages, the reservoir correction method produced widely spread age ranges compared to that of the equivalent modelled marine curve calibrated dates. However, for two archaeological sites analysed in this study with shell dates <650 years BP, the modelled marine curve calibrations agreed more closely with the archaeologists interpretation of the age of the site, than the reservoir corrected calibrated dates.

From the comparison of marine shell radiocarbon age calibration methods in Chapter 7, it was concluded that marine shell determinations calibrated using the modelled marine curve calibration method produced accurate calendar age ranges for New Zealand archaeological sites, and more precise age ranges.

## THE SAMPLE TYPE DILEMMA REVISITED

The large number of radiocarbon age determinations derived from this study has allowed the identification of reliable shellfish species for dating New Zealand prehistory. It has also supported the use of the modelled marine curve for calibrating marine shell CRA from New Zealand archaeological sites. The work of Anderson (1991) and Spriggs and Anderson (1993) highlighted the problems associated with charcoal radiocarbon chronologies from New Zealand and Polynesian archaeological sites. Their investigations allowed the rejection of dubious charcoal radiocarbon ages based on factors identified as affecting the dating of this sample type (see Chapters 3 and 4). Identifying reliable shellfish species for radiocarbon dating allows further examination of radiocarbon dates from New Zealand archaeological sites. By undertaking an analysis of all archaeological sites with paired charcoal and marine shell radiocarbon ages, shellfish species identified as being reliable in this study and application of the modelled marine curve for calibrating marine shell CRA may be further tested. In addition, the date of colonisation of New Zealand may be determined based on archaeological sites where both charcoal and marine shell radiocarbon chronologies are in agreement.

## "Chronometric Hygiene" for New Zealand Archaeological Radiocarbon Dates

The databases of the University of Waikato and IGNS Radiocarbon Dating Laboratories were searched for any archaeological sites possessing paired charcoal and marine shell dates from the same provenance (Table 8.2). To separate out acceptable charcoal/marine shell pairs from this sample, a "discard protocol" based on that developed by Anderson (1991), Spriggs and Anderson (1993) and Schmidt (1993; 1996a) was devised.

In applying a discard protocol to radiocarbon chronologies, a date is rejected firstly if it has not been measured or calculated according to international convention, and secondly, where it does not date the event being investigated. The protocol, outlined below, was used to identify spurious charcoal/marine shell pairs from New Zealand archaeological sites. Rejection or acceptance of dates was based first on the interpretation by the laboratory and researcher of the reliability and relevance of the date, and then on my own interpretation.

### Discard Protocol

*1. Charcoal radiocarbon ages may be excluded where:*

A - the charcoal has not been identified to tree species. Charcoal and wood samples used for radiocarbon dating before 1976 in New Zealand were very often not identified to tree and shrub species. Hence such radiocarbon ages on some of these samples retain the risk of high inbuilt age.

B - charcoal used for dating after 1976 is known not to have been identified;

C - the identified charcoal dated is still at risk of possessing high inbuilt age. Dates rejected on this criterion should be few in number, because of the post-1976 identifications. However, opinion is divided on the effect of particular species on charcoal $^{14}$C dates. In this discard protocol, any charcoal which has not been derived from twigs, or where the twig material is dominated by long-lived species, is rejected.

D - the charcoal dated is believed by the laboratory or researcher to be from fossil wood;

E - the sample used for dating has been noted by the laboratory or researcher to have been or was at risk of being contaminated.

*2. Marine shell radiocarbon ages may be excluded where:*

F - the species was not identified;

G - a mixture of species has been used where one or more of the species has not been identified in this study as being reliable for radiocarbon dating;

H - dates of *Amphibola crenata* (mudsnail) or fresh water shellfish, as these molluscs are known to be unreliable for dating (Anderson 1991:768; Higham 1993:122);

I - the sample used for dating has been noted by the laboratory or researcher to have been or was at risk of being contaminated;

J - the radiocarbon age has been determined from re-crystallised shell (Anderson 1991:768; Higham 1993:99);

K - the shell has come from a catchment known to be rich in ancient carbonate, which is believed by the laboratory or researcher to have affected the true age of the shell;

L - the shell has come from an environment thought to have experienced oceanic upwelling of depleted $^{14}$C, and is believed by the laboratory or researcher to have affected the true age of the shell.

*3. Radiocarbon ages may also be rejected where:*

M - Unacceptable materials have been used, or where the reliability of the material as a dating medium is at present unknown for New Zealand archaeology. These include peat and kumara.

N- the dates are far removed statistically (by more than three sigma) from other dates with the same provenance. Such dates appear obviously spurious, and even though no direct reason may account for this, they must be viewed as risky.

*4. Archaeological Relevance*

O - All samples dated must be in direct stratigraphic association with an archaeological event. Ages taken from materials defining a natural event which cannot be directly associated with the archaeological event are ignored;

P - Dates from archaeological sites where there is evidence or the possibility of post-depositional disturbance are rejected. Here the exact provenance and chronological integrity of the sample is dubious;

*5. Rejection of samples for charcoal/marine shell radiocarbon age comparisons*

Q - All radiocarbon ages where the matching charcoal or marine shell partner has been rejected, are excluded from further analysis as they have no comparative age.

*6. Acceptable New Zealand Radiocarbon Dates.*

R - Radiocarbon dates from New Zealand archaeological sites are accepted when they pass all the tests above.

## Results and Analysis

Sites with acceptable charcoal/marine shell pairs isolated from table 8.2 after application of the discard protocol, can be seen in table 8.3. Only eight sites from the original 40 remain for further analysis. The majority of sites were rejected because the charcoal used to date the archaeological deposits was unidentified. Columns 4 and 5 in table 8.3 present the acceptable pairs calibrated in years cal AD (see Chapter 2). The calibrated dates in columns 4 and 5 are given as calibrated age ranges. When age ranges from calibrated dates overlap at $1\sigma$, they are seen as being in chronological agreement (see Chapter 7). The calibrated radiocarbon dates for each site are shown in Figures 8.1 to 8.7. Marine shell ages calibrated using the Modelled Marine Curve have 'MM' appended to each calibration plot, reservoir corrected calibrated dates 'RC', and calibrated charcoal ages 'Ch'.

Calibrated charcoal ages for all sites agree closely with paired marine shell samples, calibrated by either marine calibration method (Table 8.3 and Figures 8.1 - 8.7). The shell dates from Awaroa Inlet are statistically indistinguishable ($A(p)$ = 785 ± 28 ($T$ = 0.28; $\chi^2_{1,0.05}$ = 3.84)), as are the shell ages from Shag Mouth ($A(p)$ = 995 ± 11($T$ = 8.16; $\chi^2_{10,0.05}$ = 22.36)), indicating that dates on these shellfish species are reproducible. The charcoal radiocarbon dates from Awaroa ($A(p)$ = 448 ± 38 ($T$ = 0.15; $\chi^2_{1,0.05}$ = 3.84)) and Shag Mouth ($A(p)$ = 596 ± 35 ($T$ = 0.01; $\chi^2_{1,0.05}$ = 3.84)) are also statistically indistinguishable, and illustrate that identification of the charcoal samples to twigs of short-lived species has effectively reduced inbuilt age.

| Site Name | Lab No. | Sample Type | Sample Provenance and Details | CRA | Rejection Criteria (Where applicable) |
|---|---|---|---|---|---|
| Awaroa Inlet N26/18 | Wk-1827 | Charcoal (ID) | Layer 2 | 430 ± 60 | R |
| Awaroa Inlet N26/18 | Wk-2094 | Charcoal (ID) | Layer 2 | 460 ± 50 | R |
| Awaroa Inlet N26/18 | Wk-2869 | Shell (As) | Layer 2 | 800 ± 40 | R |
| Awaroa Inlet N26/18 | Wk-2870 | Shell (Pa) | Layer 2 | 770 ± 40 | R |
| Buller Mouth S61/24 | NZ-1284 | Charcoal sp.? | Moa-hunter occupation layer | 710 ± 40 | A |
| Buller Mouth S61/24 | NZ-1285 | Charcoal sp.? | Moa-hunter occupation layer | 619 ± 40 | A |
| Buller Mouth S61/24 | NZ-1346 | Shell sp.? | Moa-hunter occupation layer | 795 ± 25 | F |
| Cook's Cove N90/644 | NZ-631 | Charcoal sp.? | Lower occupation layer | 525 ± 44 | A |
| Cook's Cove N90/644 | NZ-632 | Shell (Ls) | Lower occupation layer | 1008 ± 60 | Q |
| False Island S184/3 | NZ-144 | Charcoal sp.? | Charcoal associated with fish bone, shell and artefacts | 345 ± 50 | A |
| False Island S184/3 | NZ-145 | Charcoal sp.? | Charcoal associated with fish bone, shell and artefacts | 215 ± 50 | A |
| False Island S184/3 | NZ-141 | Shell (Pa) | Shell associated with fish bone, charcoal and artefacts | 789 ± 39 | Q |
| Foxton N148/1 | NZ-682 | Charcoal sp.? | Layer F Sq. G10 | 728 ± 42 | A |
| Foxton N148/1 | NZ-684 | Charcoal sp.? | Layer C Sq. U11 | 517 ± 60 | A |
| Foxton N148/1 | NZ-685 | Charcoal sp.? | Layer C Sq. U11 | 96 + 30 - 85 | A |
| Foxton N148/1 | NZ-683 | Shell (Cs) | Layer 1 Sq. U13 | 1053 ± 58 | Q |
| Foxton N148/1 | NZ-1250 | Shell (As) | Layer TT Sq. V14 | 895 ± 54 | Q |
| Foxton N148/1 | NZ-1251 | Shell (Cs) | Layer A Sq. U13 | 1189 ± 70 | Q |
| Foxton N148/1 | NZ-1347 | Shell (Cs) | Layer B Sq. V14 | 1159 ± 48 | Q |
| Foxton N148/1 | NZ-1349 | Shell (Cs) | Layer D Sq. T15 | 1093 ± 52 | Q |
| Foxton N148/1 | NZ-1479 | Shell (Cs) | Layer TT Sq. V14 | 984 ± 56 | Q |
| Foxton N148/1 | NZ-1480 | Shell (As) | Layer D Sq. T15 | 917 ± 55 | Q |
| Glen Gynk S61/24 | NZ-1533 | Charcoal sp. ? | Charcoal from occupational deposit | 932 ± 76 | A |
| Glen Gynk S61/24 | NZ-1532 | Shell (L, M) | Shell from occupational deposit | 568 ± 53 | G |
| Hot Water Beach N44/69 | NZ-1169 | Charcoal sp.? | Layer 4 Sq. M12 | 437 ± 44 | A |
| Hot Water Beach N44/69 | NZ-1170 | Black greasy sand sp.? | Layer 4 Sq. M11 | 492 ± 87 | A |
| Hot Water Beach N44/69 | NZ-1171 | Black greasy sand sp.? | Layer 4 Sq. M11 | 185 ± 86 | A |

| Site Name | Lab No. | Sample Type | Sample Provenance and Details | CRA | Rejection Criteria (Where applicable) |
|---|---|---|---|---|---|
| Hot Water Beach N44/69 | NZ-1296 | Shell (Pa) | Layer 4 Sq. M11 | 783 ± 40 | Q |
| Hot Water Beach N44/69 | NZ-1297 | Shell (Pa) | Layer 4 Sq. M11-12 | 854 ± 40 | Q |
| Houhora N6/4 | NZA-2436 | Charcoal (ID) | Layer 3 | 632 ± 86 | P |
| Houhora N6/4 | NZA-2438 | Charcoal (ID) | Layer 3 | 727 ± 86 | P |
| Houhora N6/4 | NZ-7920 | Shell (Cs, Pa) | Layer 3 | 812 ± 37 | Q |
| Hurumoimoi T12/347 | Wk-2853 | Charcoal (ID) | Layer 3 | MODERN | P |
| Hurumoimoi T12/347 | Wk-2854 | Shell (Pa) | Layer 3 | 670 ± 50 | P |
| Kaiwaka Pt. *Pa* N73/222 | NZ-6272 | Charcoal (ID) | Layer 1 | 268 ± 75 | R |
| Kaiwaka Pt. *Pa* N73/222 | NZ-5260 | Shell (Cs, Ps) | Layer 1 | 541 ± 32 | R |
| Kaiwaka Pt. *Pa* N73/222 | NZ-6273 | Charcoal (ID) | Layer 2 Feature B | 322 ± 76 | Q |
| Kaiwaka Pt. *Pa* N73/222 | NZ-5258 | Shell (Cs, Pa, M) | Layer 2 Feature B | 611 ± 34 | G |
| Katiki Pt. S146/4 | NZ-696 | Charcoal (ID) | House Site 2 | 880 ± 39 | C |
| Katiki Pt. S146/4 | NZ-697 | Shell sp.? | Contemporary midden | 611 ± 56 | F |
| Kaupokonui N128/3 | NZ-722 | Charcoal sp.? | Layer 2 | 1455 ± 58 | A |
| Kaupokonui N128/3 | NZ-3932 | Shell (Hi) | Layer 2 | 510 ± 50 | Q |
| Kaupokonui N128/3 | NZ-3933 | Shell (Pc) | Layer 2 | 520 ± 40 | Q |
| Manawatu N148/25 | NZ-6267 | Charcoal (ID) | Dune midden Sq Beta | 455 ± 60 | C |
| Manawatu N148/25 | NZ-5279 | Shell (Ps) | Dune midden Sq. Alpha | 807 ± 34 | Q |
| Manawatu N148/25 | NZ-5280 | Shell (Ps) | Dune midden Sq. Alpha | 834 ± 34 | Q |
| Moawhitu S10/13 | NZ-481 | Charcoal sp.? | Lower occupation layer | 674 ± 90 | A |
| Moawhitu S10/13 | NZ-482 | Shell sp.? | Lower occupation layer | 961 ± 88 | F |
| Old Pier Pt. S49/46 | NZ-3827 | Charcoal sp.? | Occupational deposit | 840 ± 60 | A |
| Old Pier Pt. S49/46 | NZ-2719 | Shell (Pr) | Occupational deposit | 1190 ± 40 | Q |
| Old Pier Pt. S49/46 | NZ-2718 | Shell (L) | Occupational deposit | 1210 ± 30 | Q |
| Oruarangi N49/28 | NZ-4177 | Charcoal sp.? | Top of mud | 230 ± 80 | A |
| Oruarangi N49/28 | NZ-4179 | Charcoal (Bracken) | Top of mud | 280 ± 50 | Q |
| Oruarangi N49/28 | NZ-4178 | Shell sp.? | Top of mud | 571 ± 40 | F |
| Orurukamatua T12/437 | Wk-2949 | Charcoal (ID) | Layer 4 | 210 ± 90 | Q |
| Orurukamatua T12/437 | Wk-2872 | Shell sp.? | Layer 4 | 690 ± 50 | F |
| Papatowai S184/? | NZ-1332 | Charcoal sp.? | Layer 4 | 916 ± 88 | A |
| Papatowai S184/? | NZ-4267 | Charcoal sp.? | Layer 4 | 854 ± 67 | A |
| Papatowai S184/? | NZ-4268 | Charcoal sp.? | Layer 4 | 759 ± 57 | A |
| Papatowai S184/? | NZ-1333 | Shell (Aa) | Layer 4 | 1381 ± 45 | Q |
| Paterangi N49/17 | NZ-4181 | Charcoal sp.? | Thin shell layer | 270 ± 80 | A |
| Paterangi N49/17 | NZ-4180 | Shell (Pa) | Thin shell layer | 580 ± 50 | Q |
| Paterson Inlet, Native Island ?/? | NZ-4193 | Charcoal sp.? | Top layer | 620 ± 30 | A |
| Paterson Inlet, Native Island ?/? | NZ-4194 | Shell (Mea) | Top layer | 600 ± 60 | Q |

| Site Name | Lab No. | Sample Type | Sample Provenance and Details | CRA | Rejection Criteria (Where applicable) |
|---|---|---|---|---|---|
| Pentland Downs S61/20 | NZ-1535 | Charcoal (ID) | Charcoal from floor deposit of limestone shelter | 1315 ± 80 | C |
| Pentland Downs S61/20 | NZ-1534 | Shell (P) | Shell from floor deposit of limestone shelter | 1240 ± 132 | Q |
| Pleasant River S155/2 | Wk-2269 | Charcoal (ID) | PLR- area 1 | 520 ± 50 | P |
| Pleasant River S155/2 | Wk-2753 | Shell (As) | PLR- area 1 | 910 ± 45 | P |
| Pleasant River S155/2 | Wk-2789 | Charcoal (ID) | PRD-7 | 550 ± 90 | P |
| Pleasant River S155/2 | Wk-2790 | Charcoal (ID) | PRD-14 | 700 ± 90 | P |
| Pleasant River S155/2 | Wk-2761 | Shell (As) | PRD-3 L4 | 1070 ± 45 | P |
| Pleasant River S155/2 | Wk-2762 | Shell (Ps) | PRD-5 L4 | 1110 ± 45 | P |
| Pleasant River S155/2 | Wk-2763 | Shell (As) | PRD-9 L3 | 920 ± 45 | P |
| Pleasant River S155/2 | Wk-2764 | Shell (Pa) | PRD-10 L3 | 790 ± 45 | P |
| Pleasant River S155/2 | Wk-2771 | Shell (Cs) | PRD-11 | 740 ± 45 | P |
| Pleasant River S155/2 | Wk-2772 | Shell (Pa) | PRD-4 | 780 ± 45 | P |
| Ponui Is. N43/1 | NZ-7764 | Charcoal (ID) | Layer D | 535 ± 55 | Q |
| Ponui Is. N43/1 | NZ-7765 | Shell (Mea) | Layer D | 957 ± 29 | G |
| Pouerua P5/195 | NZ-6980 | Charcoal (ID) | OL2 Quad 1 | 225 ± 36 | Q |
| Pouerua P5/195 | NZ-6999 | Charcoal (ID) | OL2 Quad 1 | 270 ± 50 | Q |
| Pouerua P5/195 | NZ-6893 | Shell (C) | OL3 Quad 1 | 546 ± 36 | I |
| Pouerua P5/195 | NZ-7003 | Charcoal (ID) | NII F#1 | 400 ± 40 | Q |
| Pouerua P5/195 | NZ-6885 | Shell (Hm) | OL1 F#2 | 1175 ± 68 | I |
| Pounawea S184/1 | NZ-55 | Charcoal sp.? | Intermediate layer | 500 ± 55 | A |
| Pounawea S184/1 | NZ-57 | Shell sp.? | Intermediate layer | 920 ± 55 | F |
| Pounawea S184/1 | NZ-58 | Charcoal (ID) | Lower layer | 810 ± 60 | C |
| Pounawea S184/1 | NZ-1865 | Charcoal sp.? | Lower layer | 840 ± 70 | A |
| Pounawea S184/1 | NZ-1864 | Shell (A) | Lower layer | 1110 ± 50 | Q |
| Pounawea S184/1 | NZ-1867 | Shell (A) | Lower layer | 1040 ± 50 | Q |
| Pounawea S184/1 | NZ-1868 | Shell (C) | Lower layer | 920 ± 50 | Q |
| Pounawea S184/1 | NZ-1869 | Shell (C) | Lower layer | 1040 ± 50 | Q |
| Pounawea S184/1 | NZ-1870 | Shell (A) | Lower layer | 1030 ± 50 | Q |
| Pounawea S184/1 | NZ-1871 | Shell (A) | Lower layer | 940 ± 50 | Q |
| Pounawea S184/1 | NZ-1872 | Shell (C) | Lower layer | 930 ± 50 | Q |
| Purakanui S164/18 | NZ-4637 | Charcoal (ID) | Layer 2 | 1030 ± 60 | C |
| Purakanui S164/18 | NZ-4746 | Shell (Cs) | Layer 2 | 892 ± 30 | Q |
| Purakanui S164/18 | NZ-4749 | Shell (Cs) | Layer 2 | 1020 ± 35 | Q |
| Puriri T12/340 | Wk-2641 | Charcoal (ID) | Area D2 | MODERN | R |
| Puriri T12/340 | Wk-2644 | Shell (Cs) | Area D2 | 610 ± 50 | R |
| Puriri T12/883 | Wk-2640 | Charcoal (ID) | Excavated Layer Midden F | MODERN | R |
| Puriri T12/883 | Wk-2643 | Shell (Pa) | Excavated Layer | 590 ± 50 | R |
| Puriri T12/883 | Wk-2646 | Shell (Cs, Pa) | Excavated Layer | 550 ± 45 | R |
| Puriri T12/885 | Wk-2642 | Charcoal (ID) | Excavated Layer Square H | MODERN | R |
| Puriri T12/885 | Wk-2645 | Shell (Cs) | Excavated Layer | 610 ± 45 | R |
| Redcliffs S84/76 | NZ-438 | Charcoal sp.? | Moa hunter occupational refuse | 1170 ± 60 | A |
| Redcliffs S84/76 | NZ-1111 | Shell (Cs) | Moa hunter occupational refuse | 947 ± 34 | Q |

| Site Name | Lab No. | Sample Type | Sample Provenance and Details | CRA | Rejection Criteria (Where applicable) |
|---|---|---|---|---|---|
| Sawpit Pt. N26/214 | NZ-7701i | Charcoal (ID) | L1C | 350 ± 110 | R |
| Sawpit Pt. N26/214 | Wk-3099 | Shell (Pa) | L1C | 800 ± 50 | R |
| Shag Mouth S155/5 | NZ-7761 | Charcoal (ID) | Layer 4 | 600 ± 50 | R |
| Shag Mouth S155/5 | NZ-7763 | Charcoal (ID) | Layer 4 | 592 ± 48 | R |
| Shag Mouth S155/5 | NZ-7805 | Shell (As) | Layer 4 | 965 ± 26 | R |
| Shag Mouth S155/5 | NZ-7806 | Shell (Hi) | Layer 4 | 1022 ± 29 | G |
| Shag Mouth S155/5 | NZA-1175 | Shell (Hi) | Layer 4 | 974 ± 49 | G |
| Shag Mouth S155/5 | Wk-2751 | Shell (As) | Layer 4 | 960 ± 45 | R |
| Shag Mouth S155/5 | Wk-2410 | Shell (As) | Layer 4 | 1020 ± 50 | R |
| Shag Mouth S155/5 | Wk-2362 | Shell (As) | Layer 4 | 1010 ± 50 | R |
| Shag Mouth S155/5 | Wk-2411 | Shell (As) | Layer 4 | 990 ± 45 | R |
| Shag Mouth S155/5 | Wk-2412 | Shell (As) | Layer 4 | 980 ± 45 | R |
| Shag Mouth S155/5 | Wk-2632 | Shell (Pa) | Layer 4 | 980 ± 40 | R |
| Shag Mouth S155/5 | Wk-2752 | Shell (Mea) | Layer 4 | 1040 ± 45 | G |
| Shag Mouth S155/5 | Wk-2508 | Shell (As) | Layer 4 | 1060 ± 45 | R |
| Shag Mouth S155/5 | Wk-2364 | Shell (As) | Layer 4 | 1040 ± 50 | R |
| Shag Mouth S155/5 | Wk-2856 | Shell (As) | Layer 4 | 980 ± 40 | R |
| Shag Mouth S155/5 | Wk-2857 | Shell (As) | Layer 4 | 950 ± 45 | R |
| Stewart Is. ?/? | NZ-422 | Charcoal sp.? | Layer 3 | 690 ± 60 | A |
| Stewart Is. ?/? | NZ-423 | Shell (Cs) | Layer 3 | 730 ± 50 | Q |
| Sunde site N38/24 | NZ-1899 | Charcoal sp.? | Layer 2, above ash | 620 ± 70 | A |
| Sunde site N38/24 | NZ-6955 | Shell (Pa) | Layer 2, above ash | 661 ± 36 | Q |
| Sunde site N38/24 | NZ-1898 | Charcoal sp.? | Layer 4, beneath ash | 610 ± 60 | A |
| Sunde site N38/24 | NZ-6956 | Shell (To) | Layer 4, beneath ash | 1015 ± 32 | Q |
| Tai Rua S136/1 | NZ-750 | Charcoal sp.? | Base of Layer 5 | 831 ± 33 | A |
| Tai Rua S136/1 | NZ-749 | Shell (Hi) | Base of Layer 5 | 815 ± 32 | Q |
| Tairua N44/2 | NZ-594 | Charcoal sp.? | Bed 2 | 878 ± 49 | A |
| Tairua N44/2 | NZ-595 | Charcoal sp.? | Bed 2 | 443 ± 40 | A |
| Tairua N44/2 | NZ-1875 | Shell (Cd, Ls) | Bed 2 | 910 ± 60 | G |
| Tumbledown Bay S94/30 | NZ-7656 | Charcoal (ID) | Layer 3 | 418 ± 47 | R |
| Tumbledown Bay S94/30 | NZ-7654 | Shell (As) | Layer 3 | 706 ± 50 | R |
| Whetakura T12/341 | Wk-2950 | Charcoal (ID) | Layer 2 | 320 ± 50 | Q |
| Whetakura T12/341 | Wk-2873 | Shell sp.? | Layer 2 | 770 ± 50 | F |
| Whetakura T12/341 | Wk-2951 | Charcoal (ID) | Layer 5 | 320 ± 90 | Q |
| Whetakura T12/341 | Wk-2874 | Shell sp.? | Layer 5 | 710 ± 50 | F |
| Waimataitai S146/2 | NZ-1249 | Charcoal sp.? | Moa hunter deposit | 715 ± 45 | A |
| Waimataitai S146/2 | NZ-579 | Shell sp.? | Moa hunter deposit | 940 ± 32 | F |
| Waimataitai S146/2 | NZ-580 | Shell (C. sp.) | Moa hunter deposit | 874 ± 45 | Q |

Table 8.2. New Zealand archaeological sites with paired charcoal/marine shell radiocarbon dates from the same provenance. Notes: ID indicates charcoal has been identified to tree species; sp.? no species of flora or shellfish given. Marine shell species are identified by: Pa *Paphies australis*; Ps *Paphies subtriangulatum*; As *Austrovenus stutchburyi*; Cs *Chione stutchburyi*; Aa *Amphidesma australis*; Ls *Lunella smaragda*; Ha *Haliotis australis*; Hi *Haliotis iris*; Cd *Cellana denticulata*; Mea *Mytilus edulis aoteanus*; Hm *Hyridella menziesi*; To *Thais orbita*; Pc *Perna canaliculus*; A *Amphidesma sp.*; C *Chione sp.*; Pr *Protothaca sp.*; L *Lunella sp.*; M *Mytilus sp.*

| Column No. | 1 | 2 | 3 | 4 | 5 |
|---|---|---|---|---|---|
| **Site Name** | **Lab No.** | **Sample Type** | **CRA** | **Calibrated Age Range AD 1σ (Atmospheric Curve)** | **Calibrated Age Range AD 1σ (Modelled Marine Curve)** |
| Awaroa Inlet N26/18 | Wk-1827 | Charcoal (ID) | 430 ± 60 | 1440 - 1528<br>1558 - 1631 | |
| Awaroa Inlet N26/18 | Wk-2094 | Charcoal (ID) | 460 ± 50 | 1436 - 1493<br>1603 - 1611 | |
| Awaroa Inlet N26/18 | Wk-2869 | Shell (As) | 800 ± 40 | 1439 - 1492 | 1462 - 1520 |
| Awaroa Inlet N26/18 | Wk-2870 | Shell (Pa) | 770 ± 40 | 1445 - 1522<br>1583 - 1625 | 1479 - 1545 |
| Kaiwaka Pt. *Pa* N73/222 | NZ-6272 | Charcoal (ID) | 268 ± 75 | 1641 - 1685<br>1733 - 1809<br>1930 - 1954 | |
| Kaiwaka Pt. *Pa* N73/222 | NZ-5260 | Shell (Cs, Ps) | 541 ± 32 | 1674 - 1699<br>1725 - 1778<br>1802 - 1814<br>1855 - 1855<br>1918 - 1942<br>1954 - 1954 | 1693 - 1814 |
| Puriri T12/340 | Wk-2641 | Charcoal (ID) | 144 ± 32 | 1692 - 1729<br>1822 - 1922<br>1954 - 1955 | |
| Puriri T12/340 | Wk-2644 | Shell (Cs) | 610 ± 50 | 1649 - 1678<br>1765 - 1769<br>1775 - 1805<br>1939 - 1954 | 1649 - 1703 |
| Puriri T12/883 | Wk-2640 | Charcoal (ID) | 168 ± 33 | 1682 - 1735<br>1807 - 1891<br>1909 - 1935<br>1954 - 1955 | |
| Puriri T12/883 | Wk-2643 | Shell (Pa) | 590 ± 50 | 1654 - 1684<br>1733 - 1808<br>1931 - 1954 | 1660 - 1719 |
| Puriri T12/885 | Wk-2642 | Charcoal (ID) | 177 ± 33 | 1679 - 1743<br>1749 - 1759<br>1805 - 1823<br>1827 - 1885<br>1912 - 1938<br>1954 - 1954 | |
| Puriri T12/885 | Wk-2645 | Shell (Cs) | 610 ± 45 | 1650 – 1677<br>1776 - 1805<br>1940 - 1954 | 1652 - 1700 |
| Sawpit Pt. N26/214 | NZ-7701i | Charcoal (ID) | 350 ± 110 | 1447 - 1671<br>1781 - 1799<br>1945 - 1953 | |
| Sawpit Pt. N26/214 | Wk-3099 | Shell (Pa) | 800 ± 50 | 1437 - 1495<br>1601 - 1616 | 1457 - 1527 |
| Shag Mouth S155/5 | NZ-7761 | Charcoal (ID) | 600 ± 50 | 1328 - 1350<br>1390 - 1427 | |
| Shag Mouth S155/5 | NZ-7763 | Charcoal (ID) | 592 ± 48 | 1329 - 1347<br>1392 - 1429 | |
| Shag Mouth S155/5 | NZ-7805 | Shell (As) | 965 ± 26 | 1327 - 1349<br>1391 - 1411 | 1350 - 1416 |
| Shag Mouth S155/5 | Wk-2751 | Shell (As) | 960 ± 45 | 1326 - 1353<br>1360 - 1367<br>1389 - 1416 | 1338 - 1428 |
| Shag Mouth S155/5 | Wk-2410 | Shell (As) | 1020 ± 50 | 1292 - 1333<br>1339 - 1402 | 1302 - 1397 |
| Shag Mouth S155/5 | Wk-2362 | Shell (As) | 1010 ± 50 | 1294 - 1406 | 1306 - 1403 |
| Shag Mouth S155/5 | Wk-2411 | Shell (As) | 990 ± 45 | 1301 - 1374<br>1379 - 1409 | 1319 - 1412 |

| Column No. | 1 | 2 | 3 | 4 | 5 |
|---|---|---|---|---|---|
| **Site Name** | **Lab No.** | **Sample Type** | **CRA** | **Calibrated Age Range AD 1σ (Atmospheric Curve)** | **Calibrated Age Range AD 1σ (Modelled Marine Curve)** |
| Shag Mouth S155/5 | Wk-2412 | Shell (As) | 980 ± 45 | 1304 - 1371<br>1385 - 1411 | 1324 - 1417 |
| Shag Mouth S155/5 | Wk-2632 | Shell (Pa) | 980 ± 40 | 1316 - 1370<br>1386 - 1410 | 1326 - 1415 |
| Shag Mouth S155/5 | Wk-2508 | Shell (As) | 1060 ± 45 | 1285 - 1325<br>1353 - 1358<br>1368 - 1388 | 1284 - 1335 |
| Shag Mouth S155/5 | Wk-2364 | Shell (As) | 1040 ± 50 | 1288 - 1330<br>1347 - 1393 | 1293 - 1377 |
| Shag Mouth S155/5 | Wk-2856 | Shell (As) | 980 ± 40 | 1316 - 1370<br>1386 - 1410 | 1326 - 1415 |
| Shag Mouth S155/5 | Wk-2857 | Shell (As) | 950 ± 45 | 1328 - 1350<br>1390 - 1422 | 1348 - 1433 |
| Tumbledown Bay S94/30 | NZ-7656 | Charcoal (ID) | 418 ± 47 | 1445 - 1527<br>1559 - 1631 | |
| Tumbledown Bay S94/30 | NZ-7654 | Shell (As) | 706 ± 50 | 1491 - 1604<br>1608 - 1650 | 1519 - 1654 |

Table 8.3. Acceptable pairs of charcoal and marine shell radiocarbon dates from New Zealand archaeological sites.

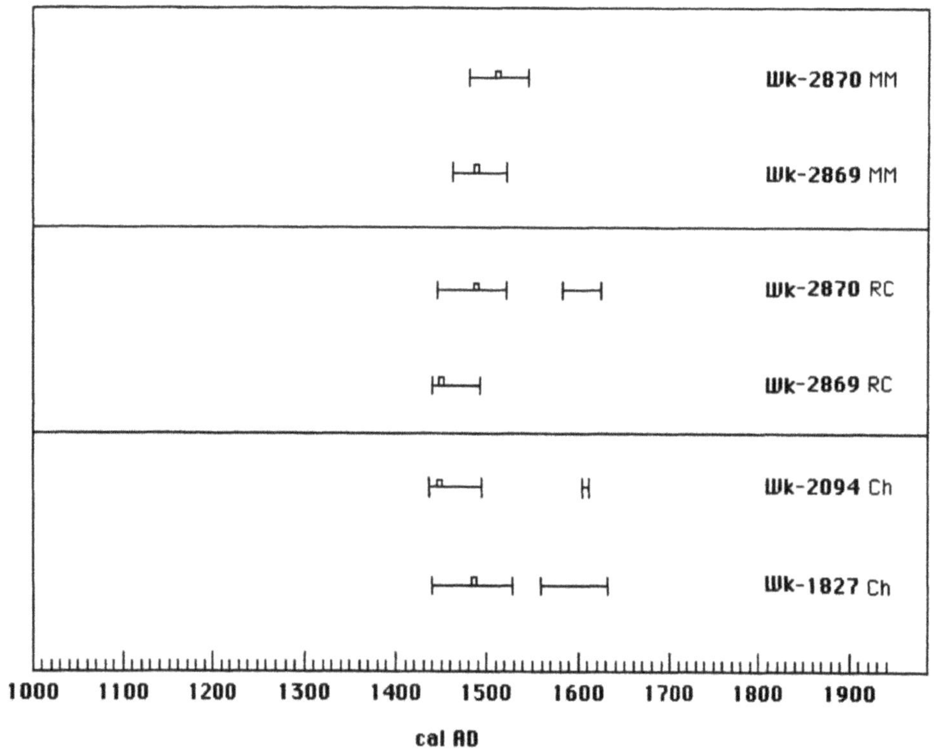


Figure 8.1. Calibrated radiocarbon ages from Awaroa (N26/18). (MM = shell calibrated by Modelled Marine Curve; RC = shell calibrated using Reservoir Correction method; Ch = calibrated charcoal age).

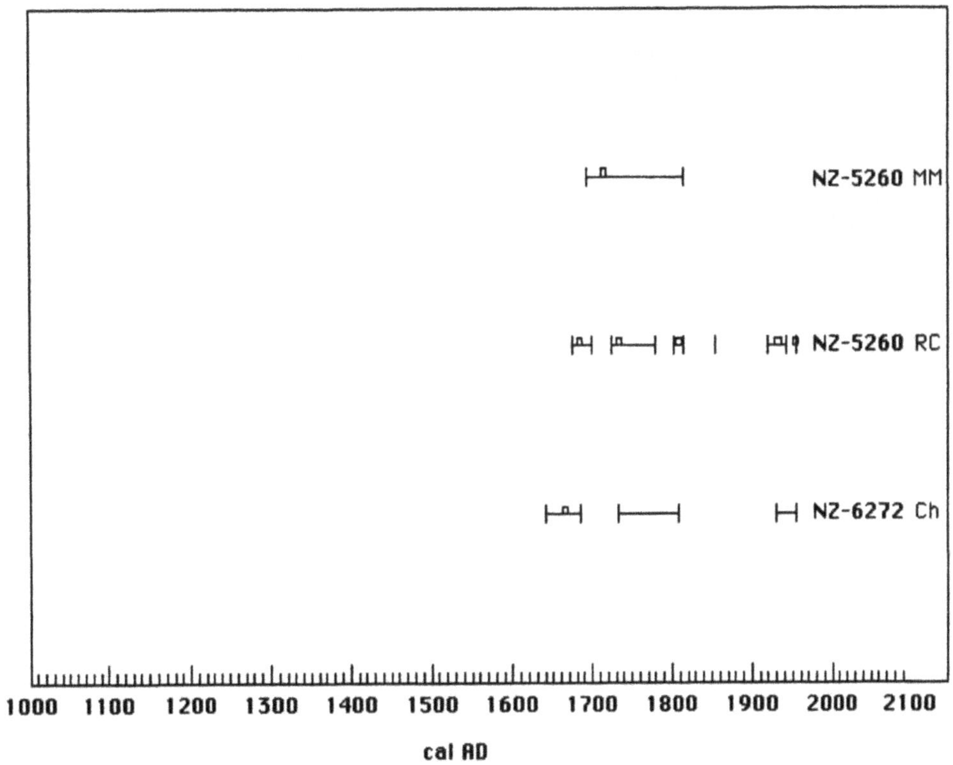


Figure 8.2. Calibrated radiocarbon ages from Kaiwaka Point *pa* (N73/322), Layer 1. (MM = shell calibrated by Modelled Marine Curve; RC = shell calibrated using Reservoir Correction method; Ch = calibrated charcoal age).

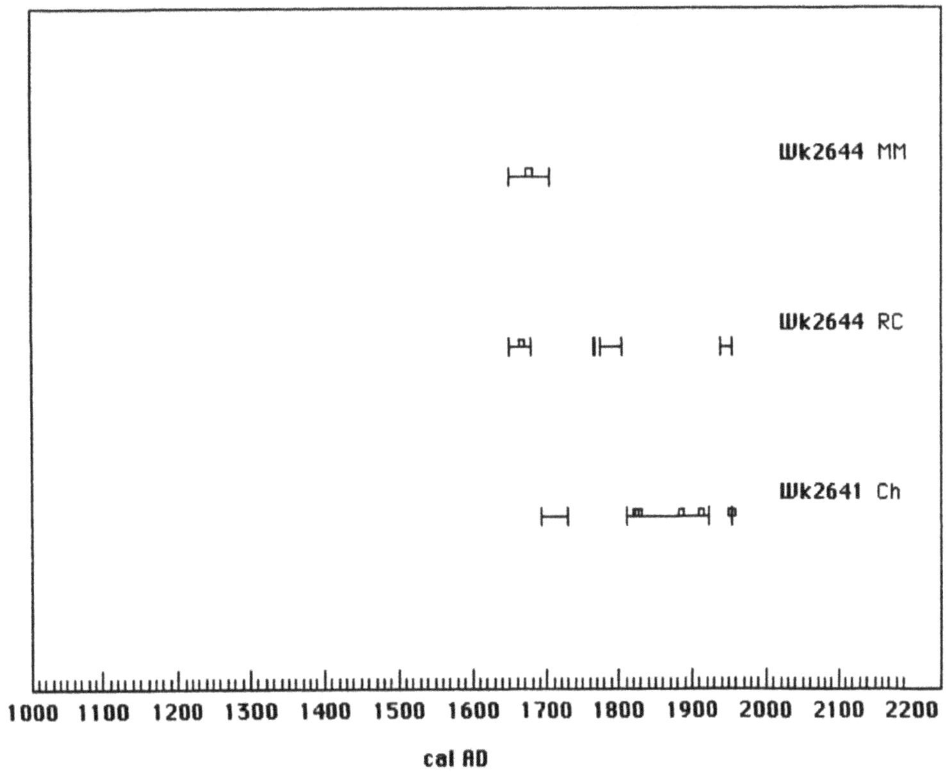


Figure 8.3. Calibrated radiocarbon ages from site T12/340, Puriri. (MM = shell calibrated by Modelled Marine Curve; RC = shell calibrated using Reservoir Correction method; Ch = calibrated charcoal age).

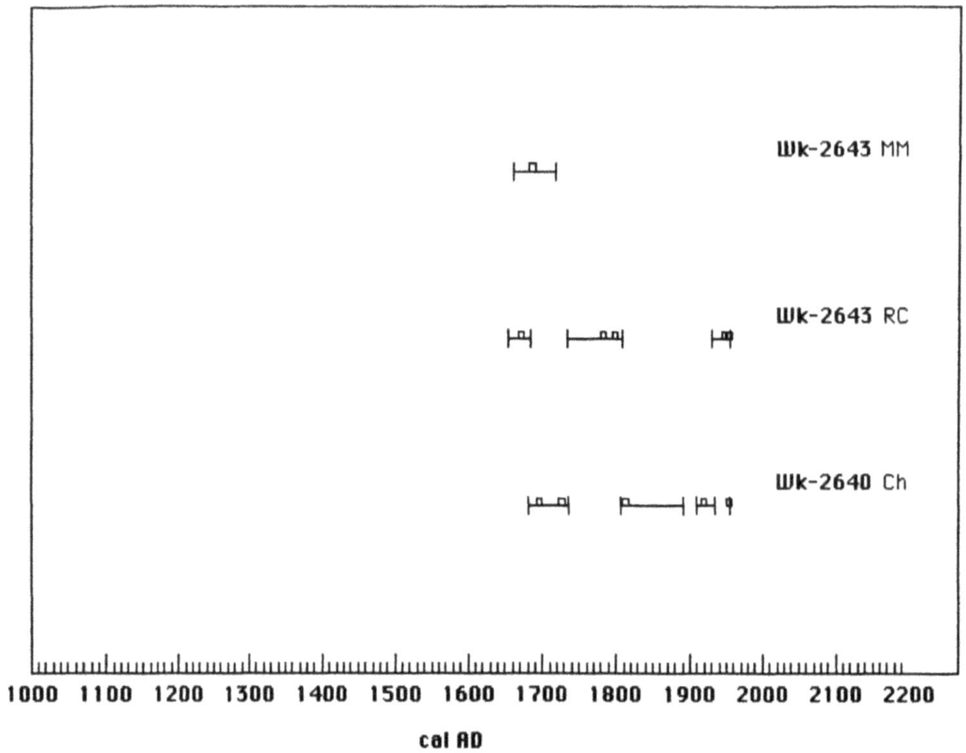


Figure 8.4. Calibrated radiocarbon ages from site T12/883, Puriri. (MM = shell calibrated by Modelled Marine Curve; RC = shell calibrated using Reservoir Correction method; Ch = calibrated charcoal age).

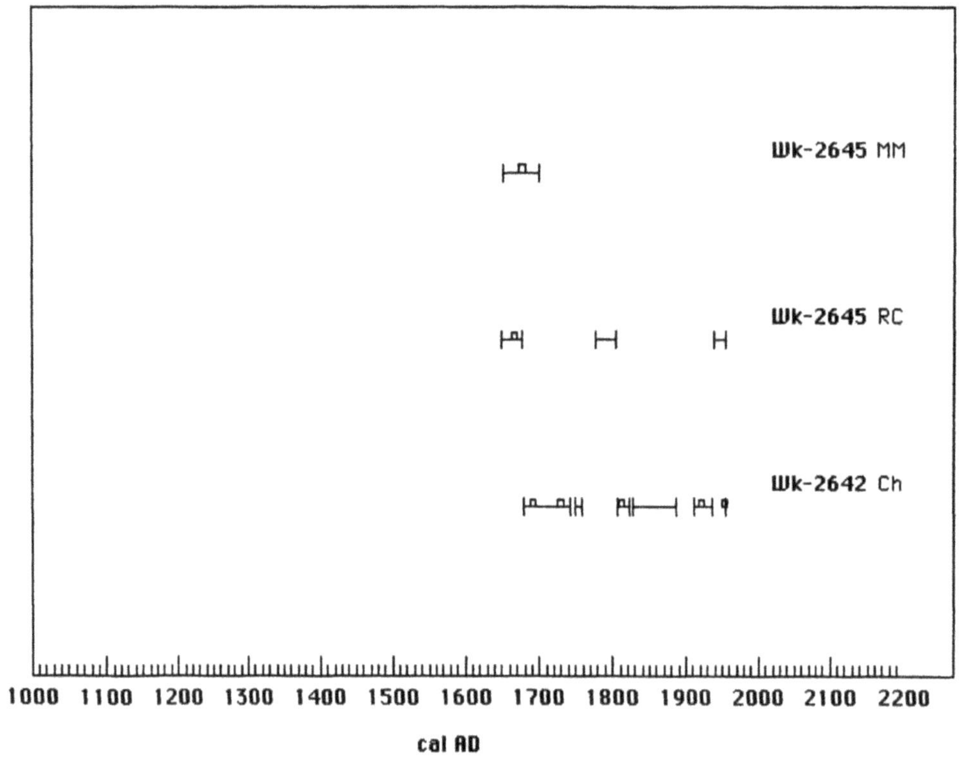


Figure 8.5. Calibrated radiocarbon ages from site T12/885, Puriri. (MM = shell calibrated by Modelled Marine Curve; RC = shell calibrated using Reservoir Correction method; Ch = calibrated charcoal age).

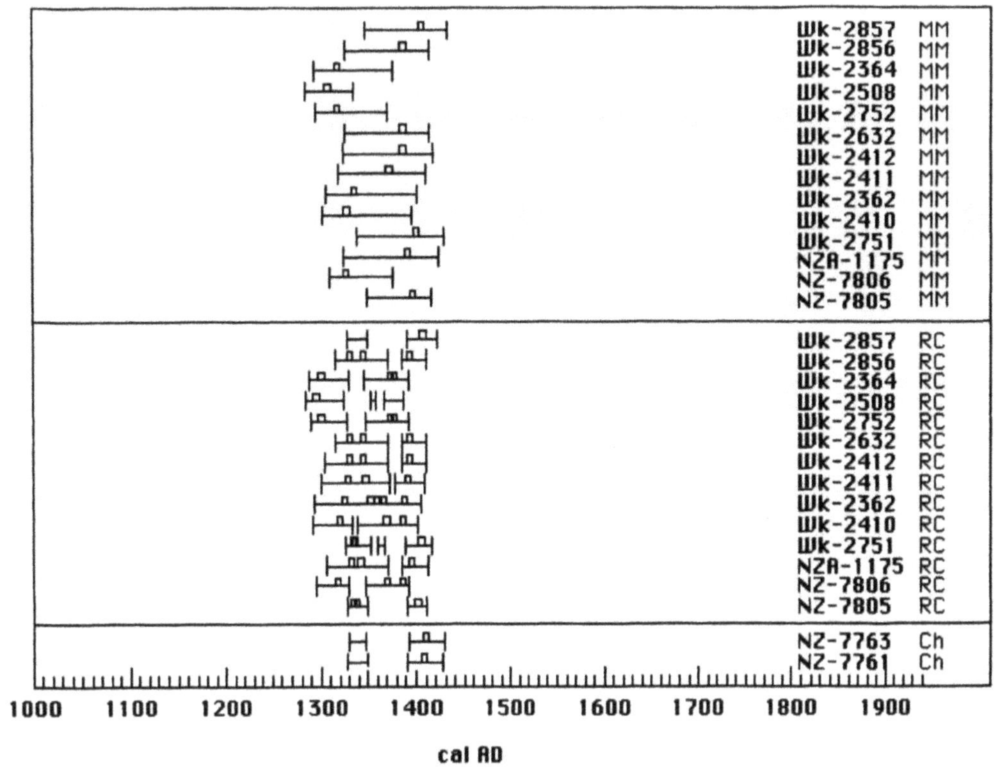


Figure 8.6. Calibrated radiocarbon ages from Shag Mouth (S155/5). (MM = shell calibrated by Modelled Marine Curve; RC = shell calibrated using Reservoir Correction method; Ch = calibrated charcoal age).

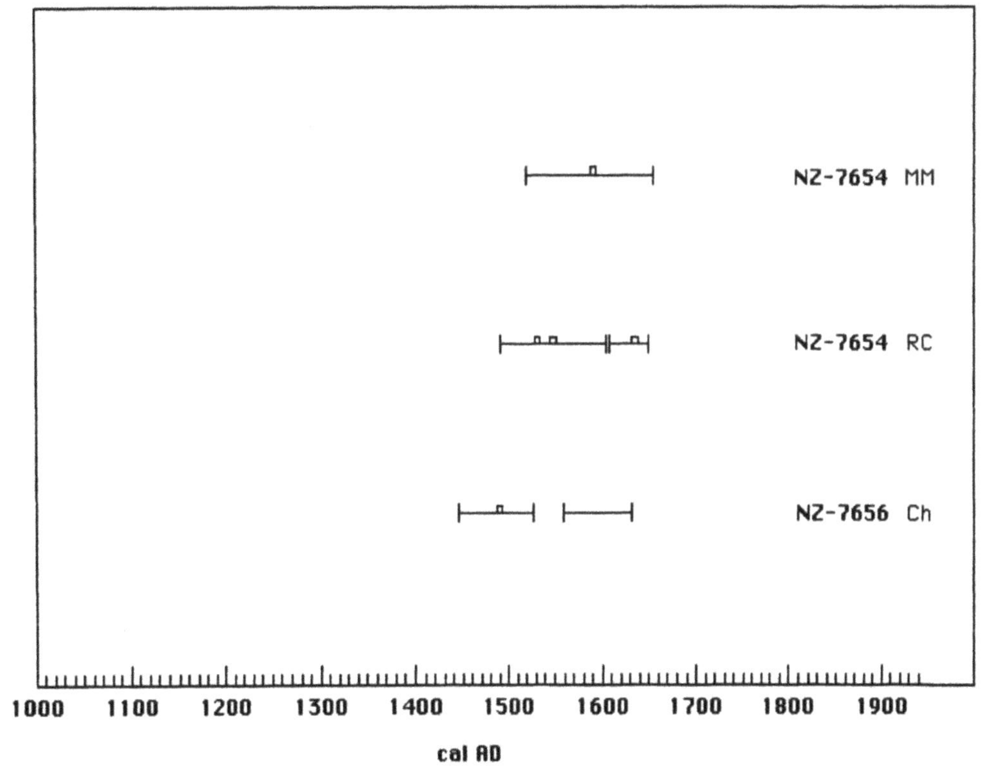


Figure 8.7. Calibrated radiocarbon ages from Tumbledown Bay (S94/30). (MM = shell calibrated by Modelled Marine Curve; RC = shell calibrated using Reservoir Correction method; Ch = calibrated charcoal age).

| Site | Pooled Mean Charcoal CRA (*A(p)*)* | Pooled Mean Shell CRA (*A(p)*)* | R(t) | ΔR |
|---|---|---|---|---|
| Awaroa Inlet (N26/18) | 448 ± 38 | 785 ± 28 | 337 ± 47 | -65 ± 47 |
| Kaiwaka Point *pa* (N73/222) | 268 ± 75 | 541 ± 32 | 273 ± 82 | -109 ± 82 |
| Shag Mouth (S155/5) | 596 ± 35 | 995 ± 11 | 399 ± 37 | 20 ± 37 |
| Tumbledown Bay (S94/30) | 418 ± 47 | 706 ± 50 | 288 ± 69 | -139 ± 69 |
| **Pooled Mean (*A(p)*)*** | | | 353 ± 25 ($T$= 3.50; $\chi^2_{3,0.05}$ = 7.81) | -39 ± 35[27] ($T$= 5.68; $\chi^2_{3,0.05}$ = 7.81) |

Table 8.4. Reservoir ages and ΔR values calculated for previously dated New Zealand archaeological sites with acceptable charcoal and marine shell pairs.* = Ward and Wilson (1978)[28].

| Source | ΔR |
|---|---|
| Higham and Hogg (1995) | -25 ± 15 |
| ΔR value from table 7.22 | -42 ± 25 |
| ΔR value from table 8.4 | -39 ± 35 |
| **Pooled Mean (*A(p)*)*** | -31 ± 15[29] ($T$= 0.40; $\chi^2_{2,0.05}$ = 5.99) |

Table 8.5. ΔR values calculated from archaeological sites analysed in this study statistically compared with the current New Zealand value from Higham and Hogg (1995). * = Ward and Wilson (1978).

This analysis of charcoal/marine shell pairs from previously dated archaeological sites, therefore, supports the results of this study in the identification of reliable shellfish species for radiocarbon dating New Zealand prehistory. This is only apparent through application of a discard protocol based on both radiocarbon and archaeological considerations.

It should be noted, that the Puriri charcoal ages are normally described by radiocarbon labs as 'MODERN' as their reservoir corrected age is less than 200 $^{14}$C years BP (see Chapter 2). However, to illustrate the differences in the two methods by which marine shell conventional radiocarbon ages may be calibrated, the actual radiocarbon dates calculated for the Puriri charcoal samples have been used.

The most notable difference between the modelled marine curve and reservoir corrected calibrated marine shell dates are the age ranges produced by each calibration technique. This was also observed in Chapter 7. Whereas the atmospheric curve can produce calibrated ages with multiple age range possibilities, the modelled marine curve provides calibrated ages where there is only one specific age range. This tendency appears to be time dependant with younger (<1500 cal AD) dates, as also concluded in Chapter 7, (see Figures 8.1 to 8.7). The charcoal and marine shell radiocarbon ages from the archaeological sites of Awaroa, Shag Mouth and Tumbledown Bay that have been calibrated on the atmospheric curve, show that their age ranges fall closely within those of the marine shell ages calibrated on the modelled marine curve. After 1500 cal AD, however, dates calibrated on the atmospheric curve become distributed over a greater time span. Though these age ranges expand with reduced radiocarbon age, marine shell dates calibrated using the modelled marine curve retain their tight chronological control. This is particularly apparent in the period from 1700 cal AD onwards for the Puriri sites and for Kaiwaka Point *pa*.

Use of the modelled marine curve for the post-1500 AD period of New Zealand prehistory is supported by the calibration of the marine shell radiocarbon ages from the Puriri sites T12/883 and T12/885, and Kaiwaka Point *pa*. At the Puriri sites, the excavator believed the sites were occupied after 1600 AD but before 1800 AD, because of an absence of European artefacts. Whereas the reservoir corrected calibrated marine shell dates indicate occupation of these sites from 1640 cal AD to the present, the modelled marine curve calibrated ages are in agreement with the excavators' estimates, showing occupation at *ca.* 1700 cal AD. At Kaiwaka Point *pa*, historic records indicate the *pa* was occupied in the 18th century (IGNS sample record form R9265/3). The shell ages calibrated using the modelled marine curve also support this occupation date. A similar

[27] Largest of weighted or unweighted 'scatter' mean σ used for ΔR rounded up to 5 years (see Stuiver *et al.* 1986:982;1993:155).

[28] The Puriri sites experience multiple ΔR values and so have been omitted from table 8.4. The charcoal and marine shell ages for Sawpit Point (N26/214) in Table 8.3 were combined with additional dates for this site in Chapter 7 to derive its ΔR value (see table 7.22), therefore no ΔR value for this site appears in Table 8.4.

[29] Largest of weighted or unweighted 'scatter' mean σ used for ΔR rounded up to 5 years (see Stuiver *et al.* 1986:982;1993:155).

vindication of the modelled marine curve based on sites believed to have been occupied after 1500 AD and prior to or just after European arrival, was observed at Area B (S8/171), Aotea, and site N43/33, Layer B, Ponui Island, in Chapter 7.

When the ΔR values are calculated for the previously dated sites, they are in agreement with the Higham and Hogg (1995) value, and the value determined for the sites sampled and analysed in this study (see Table 7.22) (Tables 8.4 and 8.5). This analysis further supports the current ΔR value for New Zealand, and the view that there has been no temporal variation in ΔR over the past 700 years for New Zealand (see Higham and Hogg 1995).

## The Chronology of Kokohuia, Ponui Island, Aotea, Ligar Bay and Sawpit Point

The earliest archaeological sites in New Zealand with acceptable paired charcoal and marine shell radiocarbon dates have been analysed in this study. Based on the agreements between the charcoal and marine shell radiocarbon chronologies, it may be proposed that New Zealand was colonised by the 14th century AD, and possibly in the 13th century AD. Events which took place at the archaeological sites sampled for this study may now be discussed within this chronological framework for New Zealand prehistory.

Taylor (1994:17) has considered Kokohuia to be an example of the late Archaic phase of New Zealand prehistory. This suggestion is given clear support by the radiocarbon results obtained (occupation beginning in the mid to late 15th century AD), the presence of a Duff 1A argillite adze, and moa bone (though moa bone associated with early sites in Northland may be at risk of being sub-fossil, see Anderson 1989:112). A full analysis of the cultural material has not been completed from Kokohuia, but the preliminary findings do shed some light on the prehistoric activities for this area of Northland.

The earliest prehistoric inhabitants at Kokohuia may have been involved in moa-hunting, though shortly after initial occupation, moa does not appear in the archaeological record. This implies either scarcity of this fauna for exploitation, local extinction, or incorporation of sub-fossil material into the deposit. The consecutive cultural deposits illustrate that the occupants depended heavily upon sea resources, by exploiting shellfish and a wide variety of fish species. Terrestrial fauna appear not to have been as relied on extensively, only two bird species being identified at present, and *Rattus exulans* and *Canis familiaris* making up the mammal component of the diet. Obsidian of possibly Mayor Island origin illustrates early access to this resource.

The radiocarbon data from Kokohuia implies occupation on this area of the ridge was short (within 100 years, and possibly within 50 years), even though there was a large amount of cultural material recovered from excavation. This brief occupation associated with extensive midden deposits is similar to that found at Shag Mouth in the South Island (Anderson *et al.* 1996). This evidence may illustrate that impact upon local resources over a short period of time at Kokohuia contributed to sudden abandonment. This site appears not to have been re-occupied until historic times.

The Kokohuia site may enable prehistorians to trace the development of Maori culture from the late Archaic Phase to the Classic Phase in Northland (Taylor 1994:17). Certainly, it could illustrate the end of moa-hunting, in association with occupation on the ridge for defensive purposes (the time of occupation of the ridge corresponding with the beginning of *pa* building in New Zealand, see Schmidt 1993, 1996a). Also, the change in foods exploited during the late Archaic (such as moa), to an increased dependence on sea resources and rat and dog in the Classic is apparent.

The dating of Area B from Whangapoua on Aotea, provides only a glimpse of prehistoric activities on this island in prehistory. Area B was occupied in the 17th century AD. The occupants depended upon marine resources, such as shellfish and fish, with apparently no exploitation of birds or mammals. Mayor Island obsidian was readily available to the site's inhabitants, as shown by its common occurrence in the site, and adzes were used and possibly manufactured. The hearths indicate that food was being prepared at that location next to the estuary, and the drain-features identified imply some structure may have existed nearby, either a house, or a storage pit.

The radiocarbon dated archaeological sites on Ponui Island, illustrate its long term occupation in New Zealand prehistory. Occupation on Ponui appears to have begun in the late 14th century AD, as indicated at Site 14[30]. Throughout the island's prehistory, there was great continuity in the subsistence activities practised by the prehistoric inhabitants. Shellfish species present today were available and exploited at most of the sites. Though shellfish appear to be the primary food source at the sampled sites, Site 14 illustrates the importance of fishing to the island's inhabitants. *Pagrus auratus* was the dominant fish species taken at this location. Shawcross (1972:594) also found *Pagrus auratus* to be the most commonly caught species in Galatea Bay to the north of Site 14. Few artefacts were found in the sites sampled for this study, but site N43/333 indicates that adzes were being used and made, and possibly clothing manufactured as seen in the presence of a bird-bone awl. The obsidian recovered from site N43/333 also shows the availability of Mayor Island and Coromandel Peninsula obsidian sources to the islands inhabitants in the 15th century AD.

The archaeological sites in Ligar Bay and at Sawpit Point were both occupied in the mid 15th century AD, and illustrate an emphasis upon local marine resources in prehistoric subsistence. Only one adze flake was found at the Ligar Bay sites, but this does indicate that adzes were being manufactured from stone resourced possibly from D'Urville Island some 75 km east. The Sawpit Point site shows that forest bird species were hunted using bird spears, though only *Callaeas cinera* was identified. Mammals exploited included *Rattus exulans*, *Canis familiaris* and *Delphinus delphis*. As found at Kokohuia, dog appears to be an important mammalian component in the prehistoric diet at this time. Prehistoric occupation from Ligar Bay and along the Granite Coast appears to have been continuous since Polynesian colonisation *ca.* the 13th century AD, as

[30] Prof. G. Irwin of the University of Auckland Department of Anthropology has recently identified an archaeological site (S11/20) on Ponui Island containing moa bone.

illustrated by the Archaic artifactual material from Wairau Bar where moa was a major part of subsistence (Duff 1950; 1956; see Challis 1991:101-142). The dating from Ligar Bay and Sawpit Point show occupation of the northern western coastline of the South Island from at least the 15th century, until the arrival of Tasman in 1642.

## CONCLUSIONS

The research presented in this dissertation leads support to the following conclusions:

1) At least nine marine shell species are reliable for radiocarbon dating New Zealand archaeological sites. These are: *Austrovenus stutchburyi, Paphies australis, Venerupis largillierti, Paphies subtriangulatum, Cominella adspersa, Cominella virgata, Crassostrea glomerata, Lunella smaragda* and *Perna canaliculus*;
2) Charcoal identified to twigs of short-lived species will provide comparative dating material to marine shell, though inbuilt age can effect reproducibility. Inbuilt age may possibly be reduced if twigs from tree and shrub species used while green by prehistoric Maori are chosen for dating;
3) Hardwater is not a potential source of error for radiocarbon dating shell from the New Zealand archaeological sites examined at the level of precision used in this study;
4) The mixing of naturally-derived shell and cultural material, and post-depositional disturbance are probable reasons for marine shell age anomalies noted by researchers in the past. However, mixing of natural and cultural material appears to be rare, and post-depositional disturbance should be able to be detected in the field by archaeologists;
5) The Stuiver and Becker (1993) modelled marine curve does provide accurate and precise calibrated ages for archaeological marine shell CRA, and is applicable for New Zealand prehistory;
6) The current $\Delta R$ value for New Zealand of -25 ± 15 years BP (Higham and Hogg 1995) is applicable for the calibration of marine shell CRA using the modelled marine curve;
7) There has been no detectable temporal variations in $\Delta R$ over the past 700 years for New Zealand;
8) New Zealand archaeological sites with acceptable paired charcoal and marine shell radiocarbon ages analysed in this study, indicate New Zealand was colonised by the 14th century AD, and possibly during the 13th century AD.

These conclusions illustrate that marine shell provides a number of advantages for New Zealand prehistorians over the use of other sample types for radiocarbon dating. The first advantage is that marine shell is ubiquitous in New Zealand archaeological sites. Second, there are no consistent offsets between radiocarbon dates on the shellfish species examined in this thesis. Third, that recrystallisation is rare in New Zealand archaeological samples. Fourth, that calibrated marine shell ages compare well with calibrated charcoal dates when calibrated using the modelled marine curve. Charcoal has the advantage of a measured terrestrial calibration curve as opposed to the modelled marine curve. However, the amount of charcoal derived from short-lived material is often very low and the possibility of inbuilt age still needs to be considered.

These advantages of using marine shell are, of course, dependent upon the nature of the archaeological site being excavated and the availability of sample types. An archaeological site dominated by charcoal which can be seen to be abundant in twigs may be of better choice for dating, especially in a region where decay of wood after death is known to be rapid. Archaeological sites such as *pa*, however, are often found to be lacking in charcoal, and marine shell therefore may be the more attractive option. In addition, the use of the modelled marine curve provides less spread in calibrated age ranges at the crucial post-1500 AD period of New Zealand prehistory.

When anomalies in radiocarbon dates from archaeological sites occur, the archaeologist has often criticised the radiocarbon dating technique (*cf.* Duff's criticism of the radiocarbon dates from Takehe Valley (Chapter 1)). I have suggested that the most common reason for age discrepancies between radiocarbon dates from archaeological sites in New Zealand is in large part due to a failure by archaeologists to take into account post-depositional disturbance. Archaeological deposits are not necessarily static, especially if an area is reoccupied. Archaeological sites which are exposed to any type of water action, such as beach fronts, or if partially submerged at high tide, should be carefully approached when sampled for radiocarbon dating. Sites which have been ploughed should also be considered cautiously.

Clearly, there are many variables which serve to influence the accuracy of radiocarbon determinations in archaeology. The validity of a radiocarbon date rests both upon the effort attached to archaeological excavation technique and to laboratory analysis. A reliable prehistoric chronology for New Zealand can only be obtained through the rigorous investigation of errors from both sources.

# BIBLIOGRAPHY


Aitken, M.J. 1990. *Science-based Dating in Archaeology.* Longman Paul, London and New York.

Allen, M. 1994. The Chronology of Coastal Morphogenesis and Human Settlement on Aitutaki, Southern Cook Islands, Polynesia. *Radiocarbon* 36(1):59-71.

Allen, J., and S. Holdaway. 1995. The contamination of Pleistocene radiocarbon determinations in Australia. *Antiquity* 69:101-112.

Ambrose, W., and Green, R.C. 1962. Obsidian Dating: Preliminary Results. *New Zealand Archaeological Association Newsletter* 5(4):247-248.

Anderson, A. 1973. The conchiolin dating method. *New Zealand Journal of Science* 16:553-558.

Anderson, A. 1982. A Review of Economic Patterns During the Archaic Phase in Southern New Zealand. *New Zealand Journal of Archaeology* 4:45-75.

Anderson, A. 1984. *The Extinction of Moa in Southern New Zealand.* In P.S. Martin and R.G. Klein (eds) *Quaternary Extinctions: A Prehistoric Revolution.* The University of Arizona Press, Tucson, Arizona.

Anderson, A. 1989. *Prodigious Birds: Moa and Moa Hunting in Prehistoric New Zealand.* Cambridge University Press.

Anderson, A. 1991. The Chronology of Colonization in New Zealand. *Antiquity* 65:767-795.

Anderson, A.J., 1996. Rat Colonisation and Voyaging: another hypothesis. *Rapa Nui Journal* 10(2):31-35.

Anderson, A., Allingham, B., and Smith, I. 1996. *Shag River Mouth: The Archaeology of an Early Southern Maori Village.* Research papers in Archaeology and Natural History No. 27, ANH Publications, The Australian National University, Canberra, Australia.

Anderson, A.J. and McGovern-Wilson, R. 1990. The Pattern Of Prehistoric Polynesian Colonisation in New Zealand. *Journal Royal Society New Zealand* 20(1):41-63.

Aravena, R., Schiff, S.L., Trumbore, S.E. and Dillion, P.J. 1991. Evaluating Dissolved Inorganic Carbon Cycling in a Forested Lake Watershed Using Carbon Isotopes. *Radiocarbon* 33(2):170.

Barber, I. 1994. *Culture Change in Northern Te Wai Pounamu.* Unpublished PhD Thesis, Department of Anthropology, University of Otago, Dunedin.

Barber, I., 1995. Constructions of Change: A History of early Maori Culture Sequences. *Journal of the Polynesian Society* 357-396.

Barnes, R.D. 1987. *Invertebrate Zoology.* CBS College Publishing, United States of America.

Barnes, R.D., Carlow, P., and Olive, P.J.W. 1993. *The Invertebrates: a new synthesis.* Blackwell Scientific Publications, Oxford.

Bellwood, P., 1973. Radiocarbon dates from Otakanini. *New Zealand Archaeological Association Newsletter* 16(4): 173-174.

Berger, R. and Meek, N. 1992. Radiocarbon Dating of *Anodonta* in the Mojave River Basin. *Radiocarbon* 34(3):578-584.

Best, E., 1914. The Peopling of New Zealand. *Man* 14:73-76.

Best, E., 1916. Maori and Maruiwi: notes on the original inhabitants of New Zealand and their culture; on the question of how that culture affected the later-coming Maori, and on the existence in these isles of customs, arts, and artefacts not traceable to Polynesia. *Transactions and Proceedings of the New Zealand Institute* 48:435-437.

Best, E., and Anderson, J.C. 1926. Another Unknown Artifact from Swamp Land. *Journal of the Polynesian Society* 35(137):70.

Beavan, N.R. and Sparks, R.J. 1998. Factors Influencing $^{14}C$ Ages of the Pacific Rat *Rattus exulans*. *Radiocarbon* 40(2): 601-613.

Bowman, S. 1990. *Interpreting the Past: Radiocarbon Dating.* British Museum Publications Ltd., London. pp. 9-15.

Brodie, J.W. 1962. A Reconnaissance of Warehou Pa, Makara. *New Zealand Archaeological Association Newsletter* 5(3):156-159.

Bronic, I.K. Horvatincic, N., Srdoc, D. and Obelic, B. 1991. Experimental Determination of the $^{14}C$ Initial Activity of Calcareous Deposits. *Radiocarbon* 33(2):181-182.

Buist, A.G. 1962. Radiocarbon Analysis. *New Zealand Archaeological Association Newsletter* 5(4):249-250.

Caughley, G. 1988. The Colonisation of New Zealand by the Polynesians. *Journal of the Royal Society of New Zealand* 18(3):245-270.

Challis, A.J., 1991. The Nelson-Marlborough Region: an Archaeological Synthesis. *New Zealand Journal of Archaeology* 13:101-142.

Challis, G.A. 1978. *Cretaceous Igneous Rocks - Tangihua Volcanics.* In R.P. Suggate (Chief Ed.) *The Geology of New Zealand, Volume II.* New Zealand Geological Survey, Department of Scientific and Industrial Research, Wellington, New Zealand. pp. 628.

Clifton, N.C. 1990. *New Zealand Timbers, Exotic and Indigenous: The Complete Guide.* Wright & Carmen Ltd., Upper Hutt.

## BIBLIOGRAPHY


Coutts, J.F. 1972. *The Utility of Radio-Carbon Determinations for Dating N.Z. Archaeological Sites.* Proceedings of the 8th International Conference on Radiocarbon Dating. pp. 516-531

Damon, P.E., Long, A. and Wallick, E.I. 1972. Dendrochronological calibration of the Carbon-14 time scale. In *Proceedings of the 8th International Conference on Radiocarbon Dating*, Lower Hutt, New Zealand. pp. 45-59.

Davidson, J.M. 1984. *The Prehistory of New Zealand.* Longman Paul, Auckland.

Deevey, E.S. (Jr) 1984. Zero BP Plus 34: 25 Years of *Radiocarbon. Radiocarbon* 26(1):1-6.

Druffel, E.M., and Suess, H.E. 1983. On the Radiocarbon Record in Banded Corals: Exchange Parameters and Net Transport of $^{14}CO_2$ Between Atmospheric and Surface Ocean. *Journal of Geophysical Research* Vol. 88, no. C2: 1271-1280.

Druffel, E.R.M. and Williams, P.M. 1991. Radiocarbon in Seawater and Organisms from the Pacific Coast of Baja California. *Radiocarbon* 33(3):291-296.

Duff, R.S. 1956. *The Moa-hunter Period of Maori Culture.* Government Printer, Wellington.

Duff, R.S. 1959. *Neolithic Adzes of Eastern Polynesia.* In J.D. Freeman and W.R. Geddes (eds). *Anthropology in the South Seas: Essays presented to H.D. Skinner.* Thomas Avery & Sons Limited, New Plymouth, New Zealand. pp. 121-147.

Duff, R.S. 1961. Excavation of House-pits at Pari Whakatau Pa, Claverly, Marlborough. *Records of the Canterbury Museum* 7(4):269-302.

Duff, R.S. 1962. Aspects of the Cultural Succession in Canterbury - Marlborough with Wider Reference to the New Zealand Area. *New Zealand Archaeological Association Newsletter* 5(4):205-209.

Dye, T. 1990. Marine Turtle Bones From an Archaeological Site in Polynesia Yield Reliable Age Determinations. *Radiocarbon* 32(2):143-147).

Dye, T. 1994. Apparent Ages of Marine Shells: Implications for Archaeological Dating in Hawaii. *Radiocarbon* 36(1):51-57.

Elzenaar, N. 1994. *Obsidian Sourcing, Whangapoua, Great Barrier Island: Areas A, B, Macrocarpa Fall & Miscellaneous Finds. 03.340 Field School, Feb. 1994.* Field School Assignment Paper 03.340, Department of Anthropology, University of Auckland, Auckland.

Facorellis, Y, Maniatis, Y. and Kromer, B. 1998. Apparent $^{14}C$ Ages of Marine Mollusk Shells from a Greek Island: Calculation of the Marine Reservoir Effect in the Aegean Sea. *Radiocarbon* 40(2): 963-973.

Fankhauser, B. 1992. *Dating Earth Ovens with Thermoluminescence and Radiocarbon.* Papers presented to the Quaternary Dating Workshop, Research School of Pacific Studies, Australian National University, Canberra, October 1990.

Fleming, C.A., and Harris, W.F. 1978. *Subdivision of the Quaternary: Palaeontological Subdivision - Hawera Series.* In R.P. Suggate (Chief Ed.) *The Geology of New Zealand, Volume II.* New Zealand Geological Survey, Department of Scientific and Industrial Research, Wellington, New Zealand. pp. 549-550.

Flessa, K.W., Cutler, A.H. and Meldahl, K.H. 1993. Time and taphonomy: quantitative estimates of time-averaging and stratigraphic disorder in a shallow marine habitat. *Paleobiology* 19(2): 266-286.

Flessa, K.W., and Kowalewski, M. 1994. Shell survival and time-averaging in near shore and shelf environments: estimates from the radiocarbon literature. *Lethaia* 27:153-165.

Fontes, J.C., Andrews, J.N., Causse, C. and Gibert, E. 1991. A Comparison of $^{14}C$ and U/Th Ages on Continental Carbonates. *Radiocarbon* 33(2):196-197.

Fox, A. 1978. *Tiromoana pa, Te Awanga, Hawke's Bay, Excavations 1974-5.* Monograph 8 of the New Zealand Archaeological Association.

Garcia, J.F., Mestres, J.S. and Rauret, G. 1991. The Use of Continental Carbonates Compared with Other Materials in the Dating of a Paleolake. *Radiocarbon* 33(2):198-199.

Gillespie, R., and Polach, H.A. 1979. *The Suitability of Marine Shells for Radiocarbon Dating Australian Prehistory.* In R. Berger and H.E. Suess (eds.) *Radiocarbon Dating*. Proceedings of the 9th International Conference on Radiocarbon Dating. University of California Press. pp. 404-421.

Glynn, P.W., Druffel, E.M. and Dunbar, R.B. 1983. A dead Central American coral reef tract: Possible link with the Little Ice Age. *Journal of Marine Research* 41:605-637.

Golson, J. 1955. Dating New Zealand's Prehistory. *Journal of the Polynesian Society* 64:113-136.

Golson, J., 1959. *Culture Change in Prehistoric New Zealand.* In J.D. Freeman and W.R. Geddes (eds). *Anthropology in the South Seas: Essays presented to H.D. Skinner.* Thomas Avery & Sons Limited, New Plymouth, New Zealand. pp. 29-74.

Golson, J., 1986. Old Guards and New Waves: Reflections on Antipodean Archaeology. *Archaeology in Oceania* 21:2-12.

Gordon, J. and Harkness, D.D. 1993. Magnitude and Geographic Variation of the Radiocarbon Content in Antarctic Marine Life: Implications for Reservoir Corrections in Radiocarbon Dating. *Quaternary Science Reviews* Vol. 11: 697-708.

Grant-Taylor, T.L. 1974. Radiocarbon Dating. *New Zealand Archaeological Association Newsletter* 17:157-162.

Green, R.C. 1962. Obsidian: Its Application to Archaeology. *New Zealand Archaeological Association Newsletter* 5(1):8-16.

Green, R.C. 1967. Sources of New Zealand's East Polynesian Culture: The Evidence of a Pearl Shell Lure

# BIBLIOGRAPHY


Shank. *Archaeology and Physical Anthropology in Oceania* 2(2): 81-90

Green, R.C. and F.W. Shawcross. 1961. Cultural Sequence in the Auckland Provenance. *New Zealand Archaeological Association Newsletter* 5(4):210-220.

Grey, C. 1994. *03.340 Report: Part I Text, Part II Illustrations. Field School Assignment Paper 03.340.* Department of Anthropology, University of Auckland, Auckland.

Grindley, G.W. 1971. *Sheet S8 - Takaka (1st ed.). Geological Map of New Zealand 1:63,3600.* Department of Scientific and Industrial Research, Wellington, New Zealand.

Grindley, G.W., 1978. *Intrusive Rocks: West Nelson.* In R.P. Suggate (Chief Ed.) *The Geology of New Zealand, Volume I.* New Zealand Geological Survey, Department of Scientific and Industrial Research, Wellington, New Zealand. pp.95-96. fig. 2.13.

Groube, L.M. 1970. *The Origin and Development of Earthwork Fortifications in the Pacific.* In R.C. Green and M. Kelly (eds) *Studies in Oceanic Culture History,* Vol. 1 PAR 11, Bishop Museum, Honolulu. pp. 133-164.

Groube, L.M., 1993. *'Dig up those moa bones, dig': Golson in New Zealand, 1954-1961.* In M. Spriggs *et al.* (eds) *A Community of Culture: The People and Prehistory of the Pacific.* Department of Prehistory, Research School of Pacific Studies, The Australian National University, Canberra, Australia. pp. 6-17.

Gupta, S.K. and Polach, H.A. 1985. *Radiocarbon Dating Practices at ANU.* Handbook, Radiocarbon Laboratory, Research School of Pacific Studies, Australian National University, Canberra.

Haast, J. 1875a. Researches and excavations carried on, in and near Moa Bone Point Cave, Sumner Road in the year 1872. *Transactions of the New Zealand Institute* 7:54-85.

Haast, J. 1875b. Notes on an ancient native burial place near the Moa Bone Point Cave, Sumner. *Transactions of the New Zealand Institute* 7:98-91.

Haast, J. 1875c. Notes on the Moa-hunter encampment at Shag Point, Otago. *Transactions of the New Zealand Institute* 7:91-98.

Hajdas, I., Beer, J., Bonani, G., Lotter, A., Sturm, M. and Wolfli, W. 1991. $^{14}$C Measurements on Laminated Lake Sediments. *Radiocarbon* 33(2):205-206.

Hamel, G. 1978. Radiocarbon dates from the moa-hunter site of Papatowai, Otago, New Zealand. *New Zealand Archaeological Association Newsletter* 21(1):53-54.

Harlow, D. 1994. *Aotea/Great Barrier Island - A History: A review of the Maori and Pakeha history of the island with a focus on Maori traditional histories, industries on Aotea and the Whangapoua area. Field School Assignment Paper 03.340.* Department of Anthropology, University of Auckland, Auckland.

Hay, R.F. 1978. *Cretaceous Stratigraphy - Mata Series.* In R.P. Suggate (Chief Ed.) *The Geology of New Zealand, Volume II.* New Zealand Geological Survey, Department of Scientific and Industrial Research, Wellington, New Zealand. pp. 367-368.

Heier-Nielsen, S.H., Heinemeier, J., Nielsen, H.L. and Rud, N. 1995. Recent Reservoir Ages for Danish Fiords and Marine Waters. *Radiocarbon* 38(2):.

Higham, T.F.G. 1993. *Radiocarbon Dating the Prehistory of New Zealand.* Unpublished PhD Thesis, University of Waikato, Hamilton.

Higham, T.F.G. 1994. Radiocarbon Dating New Zealand Prehistory With Moa Eggshell: Some Preliminary Results. *Quaternary Geochronology (Quaternary Science Reviews)* Vol. 13 pp. 163-169.

Higham, T.F.G., and Hogg, A.G. 1995. Radiocarbon dating of prehistoric shell from New Zealand and calculation of the ΔR value using fish otoliths. *Radiocarbon* 37(2): 409-416.

Hjarno, J. 1967. Maori fish-hooks in southern New Zealand. *Records of the Otago Museum (Anthropology)* No. 3.

Hogg, A. 1982. *Radiocarbon Dating at the University of Waikato, New Zealand.* Occasional Report No. 8, Department of Earth Sciences, University of Waikato, Hamilton.

Hogg, A., Polach, H., Robertson, S., and Noakes, J. 1991. *Application of High Purity Synthetic Quartz Vials to Liquid Scintillation Low-Level $^{14}$C Counting of Benzene.* In H. Ross, J.E. Noakes and J.D. Spaulding (eds), *Liquid Scintillation Counting and Organic Scintillators.* Lewis Publishers, INC. Michigan, USA.

Hogg, A.G., Lowe, D.J. and Hendy, C.H. 1987. University of Waikato Radiocarbon Dates I. *Radiocarbon* 29(2):263-301.

Holdaway, R. 1996. No Title. Paper presented on radiocarbon ages on Kiore (Rattus exulans) bone, at Department of Anthropology Seminar, University of Auckland, June 11.

Hunt, T.L. and Kirch, P.V. 1987. Radiocarbon Dates from Two Coastal Sites in the Manu'a Group, American Samoa. *Radiocarbon* 29(3): 417-419.

Ingram, B.L. 1994. *$^{14}$C Measurements in Charcoal-Shell Pairs from Indian Middens from San Francisco Bay: Paleo-Upwelling Variations Along the North American Coast.* Abstract, 15th International Radiocarbon Conference, pp. PE-11.

Ingram, B.L. and Southon, J.R. 1996. Reservoir Ages in the Eastern Pacific Coastal and Estuarine Waters. *Radiocarbon* 38(3): 573-582.

Jansen, H.S. 1984. *Institute of Nuclear Sciences INS-R-328 Radiocarbon Dating for Contributors.* Institute of Nuclear Sciences, DSIR, Lower Hutt.

Kashgarian, M., Sullivan, T.W. and Southon, J.R. 1994. *Reservoir Ages of Surface Waters in the Indonesian Seaway.*

## BIBLIOGRAPHY


Abstract, 15th International Radiocarbon Conference, pp C-10.

Kear, D. and Hay, R.F. 1961. *Sheet 1 - North Cape (1st ed.). Geological Map of New Zealand 1:250,000.* Department of Scientific and Industrial Research, Wellington, New Zealand.

Kerr, S.W. 1994. *Whangapoua - Great Barrier Island: An Analysis of Fish Bone Excavated from Sites S8/171 & S8/16. Field School Assignment Paper 03.340.* Department of Anthropology, University of Auckland, Auckland.

Kirch, P.V. and Hunt, T.L. 1988. Radiocarbon Dates from the Mussau Islands and the Lapita Colonization of the South Western Pacific. *Radiocarbon* 30(2):161-169.

Kirch, P.V., Hunt, T.L. and Tyler, J. 1989. A Radiocarbon Sequence from the Toaga Site, Ofu Island, American Samoa. *Radiocarbon* 31(1):7-13.

Law, G. 1971. Some Errors with Carbon-14 Dating. *New Zealand Archaeological Association Newsletter* 14:64-66

Law, G. 1975a. Radiocarbon Dates from Rangitoto and Motutapu, a Consideration of the Dating Accuracy. *New Zealand Archaeological Association Newsletter* 18(4):441-451.

Law, G. 1975b. C14 Dates from Harataonga Bay, Great Barrier Island. *New Zealand Archaeological Association Newsletter* 18(1):48-52.

Law, G. 1981. Radiocarbon Dating, Some Guidance for Users. *New Zealand Archaeological Association Newsletter* 24:228-236.

Law, G. 1984. *Archaeological Carbon Dating Using Marine Shell - The New Zealand Experience.* Paper presented at New Zealand Archaeological Association Conference, June 1984.

Leach, B.F. 1967. Memorandum from M. Trotter to R. Duff. Held in the Department of Anthropology, University of Otago, Dunedin.

Libby, W.F., Anderson, E.C., and Arnold, J.R. 1949. Age determination by Radiocarbon content: world-wide assay of natural radiocarbon. *Science* 109:227-28.

Little, E.A. 1993. Radiocarbon Age Calibration at Archaeological Sites of Coastal Massachusetts and Vicinity. *Journal of Archaeological Science* 20:457-471.

Lockerbie, L. 1959. From Moa-Hunter to Classic Maori in Southern New Zealand. In J.D. Freeman and W.R. Geddes (eds). *Anthropology in the South Seas: Essays presented to H.D. Skinner.* Thomas Avery & Sons Limited, New Plymouth, New Zealand. pp. 75-110.

Lockerbie, L. 1980. Correspondence between Institute of Nuclear Sciences and Les Lockerbie of the Otago Museum regarding radiocarbon dating of Pounawea archaeological site between 1957 and 1980. File reference R4544/1-9, R 9180, Institute of Geological and Nuclear Sciences, Lower Hutt, Wellington.

Long, A. 1995. From the Editor: Good-bye to Pee Dee Belemnite and Standard Mean Ocean Water; NIST Standard Nomenclature. *Radiocarbon* 37(1): iii.

Lowe, D.J., and Hogg, A.G. 1992. Application of new technology liquid scintillation spectrometry to radiocarbon dating of tephra deposits, New Zealand. *Quaternary International* 13/14:135-142.

Mangerud, J. 1972. Radiocarbon dating of marine shells including a discussion of apparent age of recent shells from Norway. *Boreas* 1:143-172.

Mangerud, J. and Gulliksen, S. 1975. Apparent Radiocarbon Ages of Recent Marine Shells from Norway, Spitsbergen, and Artic Canada. *Quaternary Research: An Interdisciplinary Journal* 5(2): 263-273.

Marchenko, E., Srdoc, D., Golubic, S., Pezdic, J. and Head, M.J. 1989. Carbon uptake in aquatic plants deduced from their natural $^{13}C$ and $^{14}C$ Content. *Radiocarbon* 31(3):785-794.

McCormac, F.G., Hogg, A.G., Higham, T.F.G., Baille, M.G.L., Palmer, J.G., Xiong, L., Pilcher, J.R., Brown, D., and Hoper, S.T. 1998. Variations of Radiocarbon in Tree Rings: Southern Hemisphere Offset Preliminary Results. *Radiocarbon* 40(3): 1153-1159.

McCulloch, B. 1973. A Relevant Radiocarbon Result (Irreverently Presented). *New Zealand Archaeological Association Newsletter* 16:128-133.

McCulloch, B and Trotter, M.M. 1975. The First Twenty Years. Radiocarbon Dates for South Island Moa-Hunter Sites, 1955-1974. *New Zealand Archaeological Association Newsletter* 18:2-17

McFadgen, B. 1972. *Paleoenvironmental studies in the Manawatu sand plain with particular reference to Foxton.* Unpublished MA thesis, Department of Anthropology, University of Otago.

McFadgen, B. 1978a. *Environment and archaeology in New Zealand.* Unpublished PhD thesis, Department of Geology, Victoria University, Wellington.

McFadgen, B. 1978b. Radiocarbon in Modern Marine Shells Compared with the New Zealand Standard. *New Zealand Journal of Science* 21:311-313

McFadgen, B. 1982. Dating New Zealand Archaeology by Radiocarbon. *New Zealand Journal of Science* 25:379-392.

McFadgen, B.G., Knox, F.B. and Cole, T.R.L. 1994. Radiocarbon Calibration Curve Variations and their Implications for the Interpretation of New Zealand Prehistory. *Radiocarbon* 36(2):221-236.

McFadgen, B.G. and Manning, M.R. 1990. Calibrating New Zealand Radiocarbon Dates of Marine Shells. *Radiocarbon* 32(2):229-232.

McGhie, M.D. 1994. *A Report on the Dating of Obsidian Artefacts from Whangapoua - Great Barrier Island, and Some Analysis of the Results. Field School Assignment Paper 03.340.* Department of Anthropology, University of Auckland, Auckland.

**BIBLIOGRAPHY**


Milligan, E.N. 1959. *The Geology of the North Hokianga District.* MSc Thesis, Department of Geology, University of Auckland, Auckland.

Moore, P.R. and Tiller, E.M. 1975. Radiocarbon Dates for New Zealand Archaeological Sites. *New Zealand Archaeological Association Newsletter* 18:98-107

Moore, P.R. and Tiller, E.M. 1976. Radiocarbon Dates for New Zealand Archaeological Sites: errata and addenda. *New Zealand Archaeological Association Newsletter* 19:151-155.

Morton, J. and Miller, M. 1968. *The New Zealand Sea Shore.* Collins, Auckland.

Oeschger, H., Siegenthaler, U., Schollerer, U. and Gugelmann, A. 1975. A box diffusion model to study the carbon dioxide exchange in nature. *Tellus* 27:168-192.

Olsson, I.U. 1983. *Dating non-terrestrial materials.* Manuscript held at University of Waikato Radiocarbon Dating Laboratory, Hamilton.

O'Regan, G., 1992. The Antiquity of the Takahe Valley Rock Shelter Site. *Archaeology in New Zealand* 35(3):172-176).

Petchey, F. 1998. *Radiocarbon Analysis of a Novel Bone Sample Type: Snapper and Barracouta Bone from New Zealand Archaeological Sites.* Unpublished PhD Thesis, University of Waikato, Hamilton.

Phelan, M.B., 1999. A ΔR Correction Value for Samoa from Known-Age Marine Shells. *Radiocarbon* 41(1): 99-101.

Polach, H.A. 1992. *Four Decades of Progress in $^{14}C$ Dating by Liquid Scintillation Counting and Spectrometry.* In R.E. Taylor, A. Long and R.S. Kra (eds) *Radiocarbon After Four Decades: An Interdisciplinary Perspective.* Springer-Verlag, New York. pp. 198-213.

Preston-Thomas, J. 1994. *S8/171 Great Barrier Island: Charcoal Analysis. Field School Assignment 03.340.* Department of Anthropology, University of Auckland.

Pullar, W.A. 1959. The significance of volcanic ash beds in archaeology. *Journal of the Polynesian Society* 68(1):8-11.

Pullar, W.A. 1959. Age of Kaharoa Ash. *New Zealand Archaeological Association Newsletter* 5(1):17-18.

Rafter, T.A., Jansen, H.S., Lockerbie, L and Trotter, M.M. 1972. *New Zealand Radiocarbon Reference Standards.* Proceedings of the 8th International Conference on Radiocarbon Dating. pp. 625-675.

Rasmussen, K.L. and Rahbek, U. 1993. *$^{14}C$ Reservoir Effect in Greenland.* Manuscript held at University of Waikato Radiocarbon Dating Laboratory.

Redvers-Newton, N.A., 1995. Radiocarbon dating human bone from Namu, Taumako, Solomon Islands; A comparison of two methods of bone pretreatment. *New Zealand Archaeological Association Newsletter* 38(2):106-116.

Redvers-Newton, N.A., and G.E. Coote, 1994. Bone pretreatments for radiocarbon dating: a study incorporating AMS dating and ion beam analysis. *Nuclear Instruments and Methods,* B29:270-273.

Riley, M. 1993. *Maori Healing and Herbal.* Viking Seven Sea Publications, Paraparaumu.

Rowland, M.J. 1976. *Cellana denticulata* in Middens on the Coromandel Coast, N.Z., - Possibilities for a Temporal Horizon. *Journal of the Royal Society of New Zealand* 6(1): 1-15.

Romanes, J.R., 1993. *Island in the sun : GIS, and a model of the prehistoric agricultural potential of Ponui Island, New Zealand.* Research essay, Department of Anthropology, University of Auckland.

Salmon, J.T. 1992. *The Native Trees of New Zealand.* Reed Books, Auckland.

Schmidt, M.D. 1993. *Few Have Been Tested by the Spade: Pa excavation and radiocarbon dating in New Zealand archaeology.* MA Research Paper, Department of Anthropology, University of Auckland, Auckland.

Schmidt, M.D. 1996a. The commencement of pa construction in New Zealand prehistory. *Journal of the Polynesian Society.*

Schmidt, M.D. 1996b. *Radiocarbon Dating New Zealand Prehistory Using Marine Shell.* Unpublished PhD Thesis, University of Waikato, Hamilton.

Schmidt, M.D and Higham, T.F.G. 1998. Origins of New Zealand's Archaic East Polynesian culture revisited: The chronology of the Tairua archaeological site. *Journal of the Polynesian Society* 107(4): 395-403.

Schofield, J.C. 1978. *Torlesse Supergroup Stratigraphy: Auckland - Northwest Waikato.* In R.P. Suggate (Chief Ed.) *The Geology of New Zealand, Volume II.* New Zealand Geological Survey, Department of Scientific and Industrial Research, Wellington, New Zealand. pp. 265.

Schofield, J.C. 1979. *Pt Sheets N38, N39, N42, N43 Waiheke. Geological Map of New Zealand 1:63,360.* Department of Scientific and Industrial Research, Wellington, New Zealand.

Shawcross, F.W. 1968. The Ngaroto Site. *New Zealand Archaeological Association Newsletter* 9(2):53-71.

Sheppard, P.J., Barker, P., Irwin, G., Jones, M., Stevenson, C., and Sutton, D.G. 1996. The development and application of New Zealand obsidian hydration dating, 1996-1998. *Archaeology in New Zealand* 39(1): 16-29.

Skinner, D.N.B. 1978. *Torlesse Supergroup Stratigraphy: Coromandel - Waikato - King Country.* In R.P. Suggate (Chief Ed.) *The Geology of New Zealand, Volume I.* New Zealand Geological Survey, Department of Scientific and Industrial Research, Wellington, New Zealand. pp. 267.

Skinner, H.D., 1921. Culture areas in New Zealand. *Journal of the Polynesian Society* 30:71-78.

Skinner, H.D., 1923. Archaeology of Canterbury. *Records of the Canterbury Museum* 3:93-104.

## BIBLIOGRAPHY


Smith, S.P., 1921. *Hawaiki: The Original Homeland of the Maori.* Christchurch: Whitcombe and Tombs.

Southon, J.R. and Kashgarian, M. 1994. *Radiocarbon Ages of Pre-Bomb Surface Ocean Waters.* Abstract, 15th International Radiocarbon Conference, pp. C-14.

Spriggs, M. and Anderson, A. 1993. Late colonization of East Polynesia. *Antiquity* 67:200-217.

Stevens, G.R. 1978. *Cretaceous Palaeontology- Mata Series.* In R.P. Suggate (Chief Ed.) *The Geology of New Zealand, Volume II.* New Zealand Geological Survey, Department of Scientific and Industrial Research, Wellington, New Zealand. pp. 362-366.

Stevenson, C.M., Sheppard, P.J., and Sutton, D.G. 1996. Advances in the Hydration Dating of New Zealand Obsidian. *Journal of Archaeological Science* 23: 233-242.

Stuiver, M. 1983. International Agreements and the Use of the New Oxalic Acid Standard. *Radiocarbon* 25(2):793-795.

Stuiver, M. and Becker, B. 1993. High-Precision Decadal Calibration of the radiocarbon Time Scale, AD 1950 – 6000 BC. *Radiocarbon* 35(1): 35-65.

Stuiver, M. and Braziunas, T.F. 1993. Modelling Atmospheric C14 Influences and C14 Ages of Marine Samples to 10,000 BC. *Radiocarbon* 35(1):137-189.

Stuiver, M., Pearson, G.W. and Braziunas, T.F. 1986. Radiocarbon Age Calibration of Marine Samples Back to 9000 CAL YR BP. *Radiocarbon* 28(2B): 980-1021.

Stuiver, M. and Polach, H.A. 1977. Discussion: Reporting of $^{14}$C Data. *Radiocarbon* 19(3):355-363.

Stuiver, M. and Reimer, P.J. 1993. Extended $^{14}$C Data Base and Revised CALIB 3.0 $^{14}$C Age Calibration Program. *Radiocarbon* 35(1):215-230.

Stuiver, M., Reimer, P.J., Bard, E., Beck, J.W., Burr, G.S., Hughen, K.A., Kromer, B., McCormac, F.G., v. d. Plicht, J., and Spurk, M., 1998a. INTCAL98 Radiocarbon age calibration 24,000 - 0 cal BP. *Radiocarbon* 40:1041-1083.

Stuiver, M., Reimer, P.J., and Braziunas, T. F. 1998b. High-precision radiocarbon age calibration for terrestrial and marine samples. *Radiocarbon* 40:1127-1151.

Suess, H.E. 1967. Bristlecone-pine calibration of radiocarbon time scale from 4100 B.C. to 1500 B.C. In *Radioactive Dating and Methods of Low-Level Counting,* IAEA, Vienna, pp. 143-151.

Suess, H.E. 1970. Bristle-cone pine calibration of radiocarbon time scale 5200 B.C. to the present. In I.U. Olsson (ed), *Radiocarbon Variation and Absolute Chronology.* Almquist and Wicksell, Stockholm, pp. 303-312.

Sutton, D.G. and Sheppard, P.J. 1994. Dating New Zealand prehistory using obsidian hydration. *Archaeology in New Zealand* 37(4): 272-281.

Tanaka, N., Monaghan, M.C. and Rye, D.M. 1986. Contribution of metabolic carbon to mollusc and barnacle carbonate. *Nature* 320:520-523.

Taylor, M. 1995. *A Preliminary Report on the Excavation at Kokohuia, South Hokianga.* Unpublished report held at University of Waikato Radiocarbon Dating Laboratory, Hamilton.

Taylor, R.E. 1987. *Radiocarbon Dating: An Archaeological Perspective.* Academic Press, Inc., San Diego, USA.

Teviotdale, D., 1932. The material culture of the moa-hunters in Murihiku. *Journal of the Polynesian Society* 41:81-120.

Teviotdate, D. 1938. Excavation of a moa-hunters camp near the mouth of the Waitaki River. *Journal of the Polynesian Society* 48:167-185.

Thompson, B.N. 1960. *Sheet 2B - Barrier (1st ed.) Geological Map of New Zealand 1:250,000.* Department of Scientific and Industrial Research, Wellington, New Zealand.

Thompson, B.N. 1978. *Stratigraphic relations of major rock groups of Northland.* In R.P. Suggate (Chief Ed.) *The Geology of New Zealand, Volume II.* New Zealand Geological Survey, Department of Scientific and Industrial Research, Wellington, New Zealand. pp. 444.

Trotter, M.M. 1967a. Radiocarbon dates from North Otago. *New Zealand Archaeological Association Newsletter* 10(4):137-150.

Trotter, M.M. 1967b. Excavations at Katiki Point, North Otago. *Records of the Canterbury Museum* 8(3):231-245.

Trotter, M.M. 1967c. Memorandum to Foss Leach. Held at the Department of Anthropology, University of Otago, Dunedin.

Trotter, M.M. 1968. On the Reliability of Charcoal for Radiocarbon Dating New Zealand Archaeological Sites. *New Zealand Archaeological Association Newsletter* 11:86-88.

Trotter, M.M. 1973. Takamatua salvage, Banks Peninsula. *New Zealand Archaeological Association Newsletter* 16(2): 76.

Trotter, M.M. and Malthus, R.S. 1967. Fluorine Analysis in New Zealand Archaeology. *New Zealand Archaeological Association Newsletter* 10(4):151-157.

Trotter, M.M. and McCulloch, B. 1984. *Moas, Men and Middens.* In P.S. Martin and R.G. Klein (eds) *Quaternary Extinctions: A Prehistoric Revolution.* The University of Arizona Press, Tucson, Arizona.

de Vires, H. 1958. Variation in concentration of radiocarbon with time and location on earth. In *Proc Koninklijke Nederlandse Akademie der Wetenschrapen,* B61, 94-102.

Vogel, J.C., Fuls, A., Visser, E. and Becker, B. 1993. Pretoria Calibration Curve for Short-Lived Samples, 1930-3350 BC. *Radiocarbon* 35(1):73-85.

Ward, G.K., and Wilson, S.R. 1978. Procedures for Comparing and Combining Radiocarbon Age Determinations: A Critique. *Archaeometry* 20(1):19-31.

Watters, M. 1978. *Tertiary Volcanism, Miocene: Great Barrier Island - Coromandel - Bay of Plenty.* In R.P. Suggate (Chief Ed.) *The Geology of New Zealand, Volume II.* New Zealand Geological Survey, Department of Scientific and Industrial Research, Wellington, New Zealand. pp. 639.

Weidman, C.R. and Jones, G.A. 1993. A Shell-Derived Time History of Bomb 14C on Georges Bank and its Labrador Sea Implications. *Journal of Geophysical Research*, Vol. 98, no. C8:14,477-14,588.

Weiner, J.S., 1955. The Piltdown Forgery. Oxford University Press, London.

Weisler, M. 1989. Chronometric Dating and Late Holocene Prehistory in the Hawaiian Islands: A Critical Review of Radiocarbon Dates from Moloka'i Island. *Radiocarbon* 31(2):121-145.

Wheeler, M. 1954. *Archaeology from the Earth.* Penguin Books, Wyman & Sons Ltd, London.

Wheeler, M. 1955. *Still Digging: Adventures in Archaeology.* Pan Books Ltd, London.

Windle, B.C.A., 1904. *Remains of the Prehistoric Age in England.* Methuen & Co., London.

Willis, E.H., Tauber H. and Munnich K.O. 1960. Variations in atmospheric radiocarbon concentration over past 1300 years. *American Journal of Science, Radiocarbon Supplement* 2:1-4.

Wilson, S.R., and Ward, G.K. 1981. Evaluation and Clustering of Radiocarbon Age Determinations: Procedures and Paradigms. *Archaeometry* 23 (1):19-39.

## BIBLIOGRAPHY

# APPENDIX 1

## Radiocarbon dating laboratories currently undertaking ΔR research

| Laboratory | Method | Region ΔR being determined for |
|---|---|---|
| Isotope and Archeometry Laboratory, University of California, U.S. | Known age shells | California seacoast and lakes |
| IGNS, Wellington, N.Z. | Known age shells | Southern North Island |
| Centre de Datation par le RadioCarbone, Universite Claude Bernard Lyon I, France | Known age shells? | Mediterranean |
| Department of Geosciences, University of Arizona, U.S. | Known age shells<br>Prehistoric charcoal/marine shell pairs | Northern Gulf of California |
| Radiocarbon Dating Laboratory, The Queen's University, Belfast, Ireland | Prehistoric bone/marine shell pairs | Ireland |
| Radiocarbon Dating Laboratory, University of Waikato, Hamilton, NZ | Prehistoric charcoal/marine shell pairs | North and South Island |
| Institute of Physics and Astronomy, University of Aarhus, Denmark (AMS) | Known age shells | Danish coastline and fiords |
| Centre for Accelerator Mass Spectrometry, Lawrence Livermore National Laboratory, California, U.S. | Known age shells;<br>Wood-shell pairs from natural beach deposits;<br>Prehistoric wood/shell pairs;<br>Varved sediments/pelagic gastropod pairs;<br>Marine/terrestrial pairs from same archaeological artefacts | California coast, east coast of United States, Mediterranean, coasts of Indian and Pacific oceans |
| IsoTrace Laboratory, University of Toronto, Ontario, Canada (AMS) | Terrestrial/marine pairs;<br>Charcoaled food/marine fur remains pairs from "Basque" copper kettles (archaeological artefact);<br>Guano and food remains/terrestrial organics pairs | North American eastern seaboard; Antarctica |

**APPENDIX 1**

# APPENDIX 2

## Radiocarbon ages measured in this dissertation

| Wk # | Location | Sample Type | Libby Age BP | $\delta^{13}C$ |
|---|---|---|---|---|
| 3463 | Aotea, Area A (S8/171) | Charcoal | 198 ± 40 | -26.0 |
| 3464 | Aotea, Area A (S8/171) | *Paphies subtriangulatum* | 720 ± 60 | 1.8 |
| 3465 | Aotea, Area A (S8/171) | *Paphies australis* | 640 ± 30 | 1.9 |
| 3466-1 | Aotea, Area A (S8/171) | *Austrovenus stutchburyi* | 510 ± 40 | 0.3 |
| 3466-2 | Aotea, Area A (S8/171) | *Austrovenus stutchburyi* | 570 ± 40 | 1.3 |
| 3467 | Aotea, Area B (S8/171) | Charcoal | 220 ± 40 | -25.7 |
| 3468 | Aotea, Area C (S8/16) | Charcoal | 220 ± 40 | -26.7 |
| 3469 | Aotea, Area C (S8/16) | *Austrovenus stutchburyi* | 820 ± 40 | 0.8 |
| 3470 | Aotea, Area C (S8/16) | *Paphies australis* | 730 ± 40 | 1.4 |
| 3471 | Aotea, Area C (S8/16) | *Paphies subtriangulatum* | 730 ± 50 | 3.4 |
| 3533 | Ligar Bay, Site N25/26 | Charcoal | 390 ± 40 | -26.3 |
| 3534 | Ligar Bay, Site N25/26 | *Paphies subtriangulatum* | 740 ± 40 | 1.2 |
| 3535 | Ligar Bay, Site N25/26 | *Paphies australis* | 760 ± 40 | 1.6 |
| 3536 | Ligar Bay, Site N25/26 | *Austrovenus stutchburyi* | 750 ± 40 | 0.1 |
| 3537 | Ligar Bay, Site N25/26 | *Lunella smaragda* | 760 ± 40 | 1.4 |
| 3538 | Ligar Bay, Site N25/26 | *Cominella adspersa* | 700 ± 40 | 0.8 |
| 3539 | Ligar Bay, Site N25/95, Midden 1 | Charcoal | 440 ± 40 | -27.6 |
| 3540 | Ligar Bay, Site N25/95, Midden 1 | *Paphies subtriangulatum* | 880 ± 40 | 1.2 |
| 3541 | Ligar Bay, Site N25/95, Midden 1 | *Paphies australis* | 820 ± 40 | 1.4 |
| 3542 | Ligar Bay, Site N25/95, Midden 1 | *Austrovenus stutchburyi* | 840 ± 40 | 0.9 |
| 3543 | Ligar Bay, Site N25/95, Midden 1 | *Lunella smaragda* | 870 ± 40 | 1.6 |
| 3544 | Ligar Bay, Site N25/95, Midden 2 | Charcoal | 470 ± 40 | -26.5 |

## APPENDIX 2

| Wk # | Location | Sample Type | Libby Age BP | $\delta^{13}C$ |
|---|---|---|---|---|
| 3545 | Ligar Bay, Site N25/95, Midden 2 | *Paphies subtriangulatum* | 790 ± 40 | 1.4 |
| 3546 | Ligar Bay, Site N25/95, Midden 2 | *Paphies australis* | 920 ± 40 | 1.6 |
| 3547 | Ligar Bay, Site N25/95, Midden 2 | *Austrovenus stutchburyi* | 840 ± 40 | 0.5 |
| 3548 | Ligar Bay, Site N25/95, Midden 2 | *Lunella smaragda* | 890 ± 40 | 1.2 |
| 3549 | Ligar Bay, Site N25/95, Midden 2 | *Mytilus edulis aoteanus* | 880 ± 40 | 2.0 |
| 3578 | Ponui Is. N43/1 | *Paphies australis* | 840 ± 40 | 1.3 |
| 3579 | Ponui Is. N43/1 | *Austrovenus stutchburyi* | 850 ± 40 | 1.0 |
| 3580 | Ponui Is. N43/1 | *Venerupis largillierti* | 860 ± 40 | 0.3 |
| 3581 | Ponui Is. N43/1 | *Perna canaliculus* | 820 ± 40 | 0.6 |
| 3582 | Ponui Is., N43/333 | Charcoal | 400 ± 40 | -25.7 |
| 3583 | Ponui Is., N43/333 | *Paphies australis* | 590 ± 50 | 1.2 |
| 3584 | Ponui Is., N43/333 | *Austrovenus stutchburyi* | 520 ± 40 | 0.8 |
| 3585 | Ponui Is., N43/333 | *Crassostrea glomerata* | 520 ± 40 | 0.8 |
| 3586 | Ponui Is. N43/333 | Charcoal | 410 ± 40 | -26.7 |
| 3587 | Ponui Is. N43/333 | Charcoal | 430 ± 40 | -26.0 |
| 3588 | Ponui Is. N43/333 | *Paphies australis* | 890 ± 40 | 0.8 |
| 3589 | Ponui Is. N43/333 | *Austrovenus stutchburyi* | 930 ± 40 | 0.7 |
| 3590 | Ponui Is. Site 14 | Charcoal | 620 ± 40 | -26.1 |
| 3591 | Ponui Is. Site 14 | *Paphies australis* | 920 ± 45 | 1.0 |
| 3592 | Ponui Is. Site 14 | *Austrovenus stutchburyi* | 940 ± 40 | 0.9 |
| 3593 | Ponui Is. Site 14 | *Perna canaliculus* | 910 ± 40 | 1.5 |
| 3594 | Ponui Is. Site 14 | *Crassostrea glomerata* | 900 ± 40 | 1.2 |
| 3596 | Ponui Is. Site 27 | Charcoal | 470 ± 40 | -26.1 |
| 3597 | Ponui Is. Site 27 | *Paphies australis* | 740 ± 40 | 1.6 |
| 3598 | Ponui Is. Site 27 | *Austrovenus stutchburyi* | 820 ± 40 | 0.5 |
| 3599 | Ponui Is. Site 95 | Charcoal | 187 ± 30 | -25.8 |
| 3600 | Ponui Is. Site 95 | *Austrovenus stutchburyi* | 770 ± 40 | 1.3 |
| 3601 | Ponui Is. Site 95 | Charcoal | 600 ± 45 | -25.0 |
| 3602-1 | Ponui Is. Site 95 | *Austrovenus stutchburyi* | 1560 ± 40 | 0.8 |
| 3602-2 | Ponui Is. Site 95 | *Austrovenus stutchburyi* | 1480 ± 50 | 0.8 |
| 3678 | Aotea, Area B (S8/171) | *Paphies subtriangulatum* | 660 ± 40 | 1.8 |
| 3679 | Aotea, Area B (S8/171) | *Paphies australis* | 680 ± 40 | 1.8 |
| 3680 | Aotea, Area B (S8/171) | *Austrovenus stutchburyi* | 630 ± 40 | 1.4 |
| 3695 | Kokohuia | Charcoal | 330 ± 45 | -25.7 |
| 3696 | Kokohuia | *Paphies australis* | 710 ± 40 | 1.14 |
| 3697 | Kokohuia | *Austrovenus stutchburyi* | 770 ± 40 | -0.1 |

## APPENDIX 2

| Wk # | Location | Sample Type | Libby Age BP | $\delta^{13}C$ |
|---|---|---|---|---|
| 3698 | Kokohuia | *Venerupis largillierti* | 680 ± 40 | 0.4 |
| 3699 | Kokohuia | *Lunella smaragda* | 760 ± 40 | 1.5 |
| 3701 | Kokohuia | *Perna canaliculus* | 800 ± 40 | 1.7 |
| 3703 | Kokohuia | Charcoal | 410 ± 40 | -26.2 |
| 3704 | Kokohuia | *Paphies australis* | 710 ± 40 | 1.0 |
| 3705 | Kokohuia | *Austrovenus stutchburyi* | 750 ± 40 | 0.1 |
| 3706 | Kokohuia | *Venerupis largillierti* | 720 ± 40 | 0.5 |
| 3707 | Kokohuia | *Lunella smaragda* | 750 ± 40 | 1.4 |
| 3709 | Kokohuia | *Perna canaliculus* | 740 ± 40 | 1.3 |
| 3711 | Kokohuia | Charcoal | 370 ± 40 | -26.2 |
| 3712 | Kokohuia | *Paphies australis* | 810 ± 40 | 1.0 |
| 3713 | Kokohuia | *Austrovenus stutchburyi* | 750 ± 40 | 0.7 |
| 3715 | Kokohuia | *Lunella smaragda* | 770 ± 40 | 1.5 |
| 3716 | Kokohuia | *Cominella virgata* | 750 ± 40 | 0.2 |
| 3718 | Kokohuia | Charcoal | 380 ± 40 | -25.7 |
| 3719 | Kokohuia | Charcoal | 440 ± 40 | -25.1 |
| 3720 | Kokohuia | Charcoal | 490 ± 40 | -26.0 |
| 3721 | Kokohuia | Charcoal | 510 ± 45 | -25.7 |
| 3722 | Kokohuia | *Paphies australis* | 790 ± 40 | 0.6 |
| 3723 | Kokohuia | *Austrovenus stutchburyi* | 810 ± 40 | -0.5 |
| 3724 | Kokohuia | *Lunella smaragda* | 820 ± 40 | 1.0 |
| 3726 | Kokohuia | *Cominella virgata* | 840 ± 45 | 0.8 |
| 3727 | Kokohuia | *Perna canaliculus* | 820 ± 45 | 1.5 |
| 4028 | Sawpit Point (N26/214) | Charcoal | 350 ± 40 | -26.8 |
| 4029 | Sawpit Point (N26/214) | Charcoal | 410 ± 40 | -26.7 |
| 4030 | Sawpit Point (N26/214) | *Austrovenus stutchburyi* | 790 ± 40 | 1.2 |
| 4031 | Sawpit Point (N26/214) | *Lunella smaragda* | 800 ± 40 | 2.1 |
| 4032 | Sawpit Point (N26/214) | *Austrovenus stutchburyi* | 1870 ± 50 | 0.9 |
| 4033 | Sawpit Point (N26/214) | *Paphies australis* | 830 ± 40 | 1.6 |
| 4482 | Kokohuia | Charcoal | 450 ± 40 | -25.5 |
| 4483 | Kokohuia | Charcoal | 490 ± 40 | -25.7 |
| 4484 | Kokohuia | Charcoal | 380 ± 40 | -26.1 |
| 4485 | Kokohuia | Charcoal | 340 ± 40 | -25.7 |
| 4486 | Kokohuia | Charcoal | 410 ± 40 | -26.0 |
| 4487 | Kokohuia | Charcoal | 380 ± 40 | -25.8 |
| 4540 | Kokohuia | Charcoal | 360 ± 40 | -25.5 |
| 4541 | Kokohuia | Charcoal | 350 ± 40 | -25.6 |

## APPENDIX 2

| Wk # | Location | Sample Type | Libby Age BP | $\delta^{13}C$ |
|---|---|---|---|---|
| 4542 | Kokohuia | Charcoal | 430 ± 40 | -25.2 |
| 4543 | Kokohuia | Charcoal | 480 ± 40 | -25.2 |
| 4544 | Kokohuia | Charcoal | 410 ± 40 | -25.2 |

www.ingramcontent.com/pod-product-compliance
Lightning Source LLC
LaVergne TN
LVHW071218210926
848482LV00015B/845